Common Well Control Hazards

Identification and Countermeasures

Common Well Control Hazards
Identification and Countermeasures

Identification and Countermeasures

By Sun Xiaozhen
Director,
Safety and Environmental Protection Supervision Center,
Xinjiang Oilfield Company

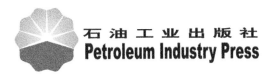

石 油 工 业 出 版 社
Petroleum Industry Press

AMSTERDAM • BOSTON • HEIDELBERG • LONDON
NEW YORK • OXFORD • PARIS • SAN DIEGO
SAN FRANCISCO • SINGAPORE • SYDNEY • TOKYO
Gulf Professional Publishing is an imprint of Elsevier

ELSEVIER

Gulf Professional Publishing is an imprint of Elsevier
225 Wyman Street, Waltham, MA 02451, USA
The Boulevard, Langford Lane, Kidlington, Oxford, OX5 1GB, UK

Library of Congress Cataloging-in-Publication Data
Sun, Xiaozhen, 1962-
    Common well control hazards : practical well control handbook : identification of hidden dangers in on-site well control equipments and remedies / by Sun Xiaozhen.
       p. cm.
    Includes bibliographical references and index.
    ISBN 978-0-12-397030-5 (alk. paper)
    1. Oil wells–Blowouts–Prevention–Handbooks, manuals, etc.   2. Gas wells–Blowouts–Prevention–Handbooks, manuals, etc.   3. Oil wells–Handbooks, manuals, etc.   4. Gas wells–Handbooks, manuals, etc.   5. Oil fields–Safety measures–Handbooks, manuals, etc.   I. Title.
    II. Title: Practical well control handbook.
    TN871.215.S86 2013
    622'.230289–dc23   2012014421

British Library Cataloguing-in-Publication Data
A catalogue record for this book is available from the British Library.

ISBN: 978-0-12-397030-5

For information on all Gulf Professional Publishing
publications visit our website at http://store.elsevier.com

Printed and bound by CPI Group (UK) Ltd, Croydon, CR0 4YY

Transferred to digital print 2012

Dedication

Qin Wengui
September 2010

Contents

Preface

I've been engaged in the work of well control technology and management since 1990. I have even participated in well control inspections organized by CNPC many times. And during the well control inspections in the national and overseas oil fields, my vision has widened and my knowledge of well control safety has increased. After communicating with peers from other oil fields on some common problems found during inspections, I got the idea to write a book.

During the well control inspections, we often encountered the following problems:

1. As a result of inconsistent understanding of the terms of the Standards, the identification of a hidden danger depends on the person who identifies it, and completely opposite conclusions are often drawn.
2. The well control hidden dangers cannot be accurately identified, and sometimes hidden dangers in well control equipment are even left out.
3. After hidden dangers are discovered, the corrective remedies and measures for them are divergent and some of them lack pertinence and feasibility.

I'd like to attempt to use safety evaluation methods to solve these problems in this book, and tell readers what problems belong to well control hidden dangers or defects, what hazards they will create, and the correct remedies and practices. Inaccuracies or mistakes are pointed out and explained, as well as how to correct them, which is why I wrote this book.

This book is presented with illustrated instructions. The contents are combined with the field practices, providing strong pertinence and feasibility. I selected more than 900 typical pictures from several thousand pictures for on-site hidden dangers collected in the last five years. Emphasis is put on the identification and analysis of 177 well control hidden dangers or defects. Based on the well site and grassroots units, this book is presented mainly with pictures supplemented by words. It represents how to identify the hidden dangers in well control equipment and points out relative remedies in plain words. I also hope that this intuitive presentation will help you become more interested in reading and will help technicians and workers on site understand and learn from this book.

There are 12 chapters in this book. Identification of well control hidden dangers and hazards and remedies for them are presented in the first 11 chapters. The problems that should be corrected are called hidden dangers. The ones that can be discussed are regarded as defects. Some foreign practices can help readers to be open-minded. Chapter 12 is focused on identifying hidden dangers pictures. The reader's ability to identify hidden dangers or defects can be developed. The answers to these dangers and defects are also provided. After having studied the first 11 chapters, readers can look at Chapter 12 as an exercise for the knowledge they have gained.

Not all the well control hidden dangers and defects on site are included in this book. The hidden dangers introduced here are only the ones that are common, typical, and ambiguous in well control inspections. Some contents are not involved in the Standards or well control textbooks, some points and ideas are not in conformity with current well control standards, and some opinions might overturn some people's traditional understanding. Everything is for your reference and discussion. There is one point I would like to make: This book is not intended to be an interpretation of the Standards. Please combine them with the actual situations in your oil field and read and make use of them selectively.

Most of the pictures in this book were taken by me on the well site during the last five years and some are provided by my peers. This book has been revised many times after I finished it. And I went to the drilling crews to ask for comments and advice. Many people gave me much help during this process. I would like to take the opportunity to express my sincere thanks to Li Dehong from Engineering Technology Company, Xiong Lasheng and Zhang Lifen from Bohai Drilling Engineering Company, Zhang Jianhua from Xibu Drilling Engineering Company, and Chen Jun from Xinjiang Training Center.

This book is suitable for drilling and well servicing operating personnel, safety supervisors, leaders of drilling crews, well control management personnel and supervisors of the organs, and drilling designers. Because of my limited knowledge, there may be unavoidable errors and inappropriate statements in the book. I will be most grateful to hear from those who detect mistakes and misleading or unclear statements and those who would like to give some advice and suggestions, all of which will be revised in the next edition.

Sun Xiaozhen
August 20, 2010, at Xinjiang Karamay Oilfield

Foreword

Sun Xiaozhen told me that he was writing a book on the identification of hidden dangers in on-site well control equipment and remedies at a conference commenting on and appraising well control inspections in 2009. I said, "Please share all your experience with us. It is helpful to increase CNPC well control management." I asked him, "How many chapters have you written?" He said that he had just begun writing. In June 2010, he gave me his script and asked me to give him some advice. He also asked me to write the foreword for this book. At first I promised to read it, but didn't promise to write the foreword. I first read the summary of each chapter, and then focused on reading the analysis on the hazards of several of the hidden dangers and defects and relative remedies. I enjoyed it so much that I told him that I was willing to write the foreword for his book.

This is a book with picture-illustrated instructions in concise language. It is written simply, and is easy to understand. Hidden dangers or defects that exist in well control equipment on site are directly pointed out in this book. It also presents the consequences that the hidden dangers will cause, and the remedies that should be adopted to remove the hidden dangers. All the pictures of the hidden dangers and remedies were taken on the well site. Every hidden danger has actually existed. One corresponding remedy is provided for each hidden danger and is feasible because all the remedies are illustrated with pictures that were taken on the well site. All the remedies are not originally created by the author, but are being put into practice or being done by our drilling crews.

The author not only knows safety technology but also has mastered well control technology and combines both in this book. The remedies in this book are highly feasible; scientificity, practicality, and economic feasibility are taken into account. Some standpoints in this book are new, and some are not included in either the Standards or well control textbooks. Many of the author's views can help to widen our perspectives. I'm gratified with the author's great innovative thought.

This book is based on the well site, and fit for the well site. It makes complex knowledge simplified using text and pictures, trying to be understandable by everyone. It is also a book that everyone will enjoy reading. This book is suitable not only for well control management personnel and technicians of all levels, but also for well control instructors and operating personnel of drilling crews.

Qin Wengui
September 2010

BRIEF INTRODUCTION TO THE AUTHOR, SUN XIAOZHEN

Sun Xiaozhen (1962–) graduated from Xinjiang Dushanzi Petroleum School in 1981, and graduated from Chengdu Institute of Geology in 1986. He is a senior drilling engineer, obtained national first-class certification of the safety evaluation division, and is a nationally certified safety engineer.

Chapter 1

Common Hidden Dangers and Remedies of Blowout Preventer (BOP) Installation

Common hidden dangers of BOP installation:

- The BOP's side holes face the drawworks.
- The pipe-ram BOP is installed above the blind-ram BOP.
- When the bell nipple tube is installed at the top of the BOP, unused bolt holes are not sealed.

Common hidden dangers of the installation of BOP manual operation rods:

- Manual operation rods are not installed completely for BOPs with a manual locking mechanism.
- Connection between the manual operation rod and the locking shaft is not tightened reliably.
- Manual operation rods and locking shaft are not connected with a cardan joint.
- Obstacles hold back the hand wheel during manual operation rod rotation.
- There is an obstacle that restricts the operation rods or manual wheels.
- Hand wheel operation rods are not connected outside of the derrick substructure.
- The inclination angle between the manual operation rod and center line of the BOP manual locking shaft is greater than 30°.
- There is no operation platform at the manual wheel or the operation platform is higher than the manual wheel.
- The manual wheel operation platform is too small, or there is no mud umbrella or armrest.
- The operation manual wheel has no outer circle.
- The operation rod manual wheel is laid vertically, and the heights of the manual wheels are not unified.
- The manual operation rod near the manual wheel side is not supported firmly, and the freedom is bigger.
- On or off circles are not marked on the hanging tag of the operation wheel.

- There is no counting device when opening or closing the manual wheel.
- After shutting-in manually or locking, the wheel is turned back by a quarter- or half-circle again.

Common hidden dangers for fixing the BOP stack:

- Guy lines for fixing the BOP stack tie up the lower part of the BOP or wind around the BOP body.
- Guy lines are not anchored to the BOP.
- Guy lines are not used with a steamboat ratchet.
- Guy lines use a chain fall.
- The guy line's diameter is less than 16 mm.
- Guy lines are not laid along the diagonal of the derrick substructure.
- Guy lines (props) are tied up by coping bolts, side door bolts, or flange bolts of the BOP.
- Guy lines are fixed upward horizontally.
- Wire rope is used in tying a knot, rope clamps are not used, and rope clamps are stuck reversely or the clamp interval is not up to standard.
- Props are used to fix the BOP stack.
- Wire rope is wound around the BOP flange.

Mud umbrella fence and other common hidden dangers:

- The mud umbrella appears as a horizontal type or umbrella type.
- The mud umbrella area is too small.
- A plastic cloth instead of a mud umbrella is wound around the BOP.
- The square well (cellar) lacks an operation platform.

1. BLOWOUT PREVENTER INSTALLATION

Hidden danger: The BOP's side holes face the drawworks

Hazard

After shutting-in, the kill line passageway is blocked, so there is no way to kill the well operation. In this case, the side holes can be connected to the choke manifold or kill line with pipe, providing a new passageway for killing the well operation.

When the BOP's side holes face the drawworks (Fig. 1-1-3), it is inconvenient to connect pipe from the side holes. Reinforcing ribs of the substructure, relief line, manual locking rod, and so on will all make the stretched out pipe space restricted. Many bent subs need to be connected to the kill line and choke manifold. Otherwise, the space between the BOP and drawworks is narrow, making it inconvenient and unsafe for operators to come in or operate, as shown in Figures 1-1-1 and 1-1-2. In Figure 1-1-4, hydraulic pipe interface of the BOP faces the V-door, and the side hole faces the drawworks.

FIG. 1-1-1   Space between the BOP and draw-works is narrow (A).

FIG. 1-1-2   Space between the BOP and draw-works is narrow (B).

FIG. 1-1-3   Side hole faces the drawworks.

FIG. 1-1-4   Hydraulic pipe interface faces the V-door.

Remedy

When installing the BOP, the side hole of the BOP should face the V-door. Thus, connection has a large operation space, operators are safe, and there is no or little need to use it with bent subs. In this premise, a hydraulic pipe interface should be used, and should face the V-door. The maximum probability of leakage will occur at the hydraulic interface. Perform a daily inspection to observe any leakage, and maintain it in time more conveniently when the interface faces the V-door.

There is a large area in front of and behind the gate of the electric drilling rig. If the BOP side hole faces the V-door, the hydraulic pipe not only can face the V-door, but it can back to the V-door. In a standardization installation, however, the hydraulic pipe should face the V-door. For mechanical drilling rigs, the hydraulic pipe should face the V-door. If the hydraulic pipe interface faces the drawworks, it is difficult to observe the hydraulic pipe interface sealing circumstances behind the BOP. Especially in lower substructure MDRs, no one is willing to get into the BOP back, which is a narrow and dirty passageway that is blocked, to inspect hydraulic pipe sealing circumstances (shown in Figs. 1-1-7, 1-1-8). Roomy space on an electric drilling rig is illustrated in Figures 1-1-5 and 1-1-6.

The BOP hydraulic pipe interface should face the drawworks, as regulated in *SY/T 5964—2006 Standard*, however the side hole direction is not regulated.

At present, four methods of laying out side holes and hydraulic pipe interface are available. The recommended installation directions are as follows:

1. For a single side hole BOP where the side hole has no hydraulic pipe, the BOP side hole should face the V-door. The hydraulic pipe interface faces the drawworks, as shown in Figures 1-1-9 and 1-1-10.

FIG. 1-1-5 Roomy electrical drilling rig sub-structure (A).

FIG. 1-1-6 Roomy electric drilling rig sub-structure (B).

FIG. 1-1-7 Passageway blocked the BOP back.

FIG. 1-1-8 Space of the BOP back is narrow.

FIG. 1-1-9   Side hole faces the V-door (A).   **FIG. 1-1-10**   Side hole faces the V-door (B).

2. For a single side hole BOP where the side hole has a hydraulic pipe interface, the side hole and hydraulic pipe should face the V-door, as shown in Figures 1-1-11 and 1-1-12. In Figure 1-1-11, the BOP has hydraulic pipe interfaces on two sides. In Figure 1-1-12, there is a hydraulic pipe interface at only one side of the side hole.

FIG. 1-1-11   Side holes and hydraulic pipe interface face the V-door (A).   **FIG. 1-1-12**   Side holes and hydraulic pipe interface face the V-door (B).

3. For a BOP that has side holes on two sides, its hydraulic pipe interface should face the V-door, as shown in Figure 1-1-13. In Figure 1-1-14, there are all side holes and hydraulic pipe interfaces; therefore any side of the BOP facing the V-door is allowed. Generally, the nameplate faces the V-door.

FIG. 1-1-13 BOP has side holes at two sides.   **FIG. 1-1-14** BOP has side holes and hydraulic pipe interfaces at two sides.

4. Generally, there is no side hole for operating a BOP whose diameter (below 180 mm) and volume are small. During deepening and sidetracking in old boreholes, the hydraulic pipe interface should face the V-door, as shown in Figure 1-1-15. Single ram BOPs used in some wells have no side hole; for this type of BOP, in order to observe and deal with leakage of hydraulic pipe conveniently, the hydraulic pipe interface should face the V-door, as shown in Figure 1-1-16.

FIG. 1-1-15 Small open diameter and no side hole BOP.   **FIG. 1-1-16** No side hole BOP.

Side holes of BOPs are blinded in China and generally are not used; however overseas they are used daily. In Figure 1-1-17, valves and hydraulic pipe interfaces are connected to side holes beforehand. In Figure 1-1-18, not using a drilling spool, the relief line is connected from two sides of ram BOP to the choke manifold and kill line, respectively.

The hydraulic pipe interface of an annular BOP should face the V-door. At present, most annular BOP hydraulic pipe interfaces all face the drawworks. As such, it is difficult to observe and deal with the interface leakage circumstances.

In Figure 1-1-19, the hydraulic pipe interface faces the drawworks; there is leakage at the interface in the picture. In Figure 1-1-20, the hydraulic pipe interface faces the V-door.

FIG. 1-1-17 Side holes are connected to valves and pipe subs.

FIG. 1-1-18 Side holes are connected with a relief line.

FIG. 1-1-19 Hydraulic pipe interface leaks oil.

FIG. 1-1-20 Hydraulic pipe interface faces the V-door.

The hydraulic pipe interface of BOP should face the drawworks, which is regulated in *SY/T 5964—2006*. Some of my views, mentioned earlier, are not unified with the Standard.

Hidden danger: The pipe-ram BOP is above the blind-ram BOP

Hazard

The pipe-ram BOP is installed above the blind-ram BOP, as in Figure 1-1-21. After closing the pipe-ram BOP, there is drill string in the borehole. If the ram is stabbed and there is no way to replace it, it will cause an out-of-control blowout. A blowout happened in the SH1025 well, at the Xinjiang Oilfield, on June 30,

1995. The pipe-ram BOP was stabbed after shutting-in. Because the pipe-ram BOP was above the blind-ram BOP, there was no way to replace the ram, leading to an out-of-control blowout.

Remedy

The relative position between a pipe-ram BOP and a blind-ram BOP is not regulated definitively in the Standard. At present, there is two points of view: One considers that the blind-ram BOP should be on top; another thinks the pipe-ram should be on top. During drilling, there are only two shut-in operating modes: an empty borehole or the drill string being in the borehole. Analyses of these two modes are as follows:

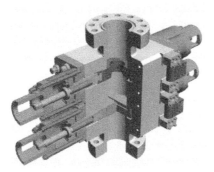

FIG. 1-1-21   Pipe-ram BOP is above the blind-ram BOP.

FIG. 1-1-22   Blind-ram BOP is above the pipe-ram BOP.

1. When the borehole is empty, close the blind-ram BOP. If the pipe-ram BOP is above the blind-ram BOP, when the blind-ram BOP is stabbed, the pipe-ram BOP can be changed into a blind-ram BOP, recovering to control the wellhead. If the blind-ram BOP is above the pipe-ram BOP, there is no way to replace the ram, and the wellhead is out of control. So, when there is an empty hole, the pipe-ram BOP should be above the blind-ram BOP, as shown in Figure 1-1-21.
2. When a drill string is in the borehole, close the pipe-ram BOP. If the blind-ram BOP is above the pipe-ram BOP and the pipe-ram BOP is stabbed, the blind-ram BOP can be changed into a pipe-ram BOP, recovering to control the wellhead. If the pipe-ram BOP is above the blind-ram BOP, there is no way to replace the ram, so the wellhead is out of control. Therefore, when a drill string is in the borehole, the blind-ram BOP should be above the pipe-ram BOP, as shown in Figure 1-1-22.

Most blowouts happen in drilling mode or tripping mode. Even if a blowout happens in an empty hole, the hole should be filled quickly with some drill strings.

It is regulated in the Standard that "after pulling out of hole, running in hole in time. It is forbidden to maintain equipment during empty hole." So, after gushing, there is maximum probability to use a pipe-ram BOP to shut-in the well. In *China Petroleum Group Company Blowout Disaster Case Collection,* 56 blowout cases are documented; three case disasters happened during a logging operation, and blind-ram BOP should have been used. In all the other cases, there are drill strings in the borehole, so a blind-ram BOP should be above the pipe-ram BOP, as shown in Figure 1-1-22.

Hidden danger: When the bell nipple is installed at the top of the BOP, unused thread holes are not blinded

Hazard

There are bolt holes at the top of the BOP. When the bell nipple is installed, the unused holes are not blinded. If the drain water and drilling fluid leak from the derrick floor and go into the threaded holes, it will erode and damage the threads in the holes. There is no way to connect a rotary BOP at the top of the BOP. Some drilling crews use grease to seal the holes, but upper drain water and drilling fluids still flow into the threaded holes, eroding and damaging inside threads.

Remedy

For the bell nipple installed at the top of either an annular or ram BOP, four bolts are needed to meet installation requirements. Others should be sealed with bolts or special rubber plugs (Figs. 1-1-23, 1-1-24). It is simple and convenient to seal the holes with rubber plugs instead of bolts.

FIG. 1-1-23 Unused threaded holes are sealed with rubber plugs.

FIG. 1-1-24 Unused threaded holes are sealed with bolts.

2. BOP MANUAL LEVER INSTALLATION

Hidden danger: Manual levers are not completely installed for BOPs with manual lock devices

Hazard

The manual lever has two functions: closing the well by using the rod once the hydraulic BOP doesn't work, and locking the ram BOP manually after closing the well hydraulically. If there is no manual lever installed or it is not installed completely (Figs. 1-2-1, 1-2-2), the aforementioned two functions cannot be achieved. Especially when the hydraulic pressure fails to work, the well cannot be closed manually in time.

FIG. 1-2-1 Manual lever/rod is not completely installed.

FIG. 1-2-2 No manual lever is installed.

Remedy

A BOP with a manual lock device should be installed completelywith the manual levers so that if the hydraulic cannot be used, wells can be manually shut-in and the rams can be locked in time after hydraulically shutting-in the well (Figs. 1-2-3, 1-2-4).

FIG. 1-2-3 Completely installed manual lever (A).

FIG. 1-2-4 Completely installed manual lever (B).

Hidden danger: The manual lever and locking shaft are not firmly connected

Hazard

Firm connections cannot be realized if there is no jack screw or the jack screw doesn't lock well at the connection point between the manual lever and locking shaft (Fig. 1-2-6). The lever sways when turning the hand wheel; the BOP shakes when drilling and hydraulically shutting-in the well, which causes the manual lever to drop from the locking shaft (Fig. 1-2-5).

Remedy

FIG. 1-2-5 Manual lever drops from the locking shaft while the well is shut in.

FIG. 1-2-6 Manual lever is fixed by an iron wire.

Tighten the jackscrew and manually fasten the locking shaft (Figs. 1-2-7, 1-2-8).

FIG. 1-2-7 Tighten the jackscrew (A).

FIG. 1-2-8 Tighten the jackscrew (B).

Hidden danger: A cardan joint is not used for the connection of the manual lever and locking shaft

Hazard

When a slight angle is needed for a lever to connect to the derrick substructure, if we do not use a cardan joint, it will be hard to change the angle of the lever (Figs. 1-2-9, 1-2-10).

FIG. 1-2-9   Cardan joint is not used (A).

FIG. 1-2-10   Cardan joint is not used (B).

Remedy

Using a cardan joint to connect the manual lever and locking shaft can make the lever connect to the derrick substructure with a slight angle (not over 30°); moreover, the turning lever is flexible (Figs. 1-2-11, 1-2-12).

FIG. 1-2-11   Using a cardan joint (A).

FIG. 1-2-12   Using a cardan joint (B).

Hidden danger: There is a barrier obstructing the manual lever turning scope

Hazard

A barrier within the turning scope of the hand wheel will obstruct it (Figs. 1-2-13 to 1-2-16).

FIG. 1-2-13   Hand wheel cannot turn (A).

FIG. 1-2-14   Hand wheel cannot turn (B).

FIG. 1-2-15   Barrier obstructs the turning of the hand wheel (A).

FIG. 1-2-16   Barrier obstructs the turning of the hand wheel (B).

Remedy

There should not be a barrier within the turning scope of the hand wheel, and there should be enough space for the hand wheel to turn flexibly (Figs. 1-2-17, 1-2-18).

FIG. 1-2-17   Hand wheel turns freely (A).

FIG. 1-2-18   Hand wheel turns freely (B).

Hidden danger: There is a barrier on the route of the manual lever or hand wheel

Hazard

If there is a barrier on the route of the manual lever or hand wheel, there is no way to close the lever and to lock the ram. In Figures 1-2-19 and 1-2-20, the ram is locked by the locking shaft retracting.

FIG. 1-2-19   Link rings limit the route of the rod.

FIG. 1-2-20   Stand limits the route of the lever.

Remedy

There should not be a barrier at the hand wheel that may limit the rod's route (Figs. 1-2-21, 1-2-22).

FIG. 1-2-21   The route of the manual lever is not limited (A).

FIG. 1-2-22   The route of the manual lever is not limited (B).

Hidden danger: The manual lever does not connect outside of the derrick substructure

Hazard

If the manual lever is not connected outside of the derrick substructure, the crew would need to go into the substructure to turn the hand wheel. When a kick or blowout occurs, it is very dangerous for the crew to go into the substructure because it is narrow, the floor is wet, sight is unclear, flammable gas fills the air, and the escape way is obstructed (Figs. 1-2-23, 1-2-24).

FIG. 1-2-23   Manual lever does not connect out of the derrick substructure (A).

FIG. 1-2-24   Manual lever does not connect out of the derrick substructure (B).

Remedy 1

The manual lever should connect outside of the derrick substructure; as a result, the crew can shut-in the well manually or lock outside of the substructure. In order to connect the manual lever outside of the derrick substructure while the hand wheel turns freely, we can use one or two cardan joints in the lever (Figs. 1-2-25, 1-2-26).

FIG. 1-2-25   Using two cardan joints (A).

FIG. 1-2-26   Using two cardan joints (B).

Remedy 2

If the manual lever-to-borehole distance is 15 m, the hand wheel turns freely. Set a block fire cover at the hand wheel so that even if a blowout occurs, the crew can shut-in the well manually (Figs. 1-2-27, 1-2-28).

FIG. 1-2-27 Set a block fire cover at the hand wheel (A).

FIG. 1-2-28 Set a block fire cover at the hand wheel (B).

Hidden danger: The offset angle between the manual lever and the BOP manual locking shaft center line is over 30°

Hazard

If the offset angle between the manual lever and the BOP manual locking shaft center line is over 30°, the lever will not turn freely (Figs. 1-2-29, 1-2-30).

FIG. 1-2-29 Offset angle of the lever is over 30° (A).

FIG. 1-2-30 Offset angle of the lever is over 30° (B).

Remedy

To keep the manual lever turning freely, the offset angle between the manual lever and the BOP manual locking axis center line should not be over 30° (Figs. 1-2-31, 1-2-32). However, if the offset angle is larger than 30° but the manual lever turns freely, it is also acceptable.

The manual lever's offset angle is one of the standards to judge whether or not the installation of the lever is qualified. The best way to judge the manual lever's installation eligibility is if the lever can be turned freely by two men.

FIG. 1-2-31 Offset angle of the manual lever is not over 30° (A).

FIG. 1-2-32 Offset angle of the manual lever is not over 30° (B).

Hidden danger: There is no operating platform at the manual lever hand wheel or the operating platform is too high from the hand wheel

Hazard

If the hand wheel's height from the ground is greater than the operator's height, he will not be able to turn the hand wheel because there is no operating platform (Figs. 1-2-33, 1-2-34).

FIG. 1-2-33 Lack of operating platform.

FIG. 1-2-34 The operating platform is too high above the hand wheel.

Remedy

When the hand wheel is too high to reach, there should be an operating platform at the hand wheel. To accommodate different operators' heights, the distance between the operating platform and lever should not be greater than 1.6 m; otherwise the crew cannot work conveniently (Figs. 1-2-35, 1-2-36).

FIG. 1-2-35 Hand wheel operating platform (A).

FIG. 1-2-36 Hand wheel operating platform (B).

Hidden danger: The manual lever hand wheel operating platform is too small, or it lacks a guardrail or handrail

Hazard

Strong force is needed when closing and locking ram BOP manually; as a result, at least two operators are needed to turn the hand wheel. If the operating platform is too small, two operators cannot stand on it at the same time to turn the hand wheel (Figs. 1-2-37, 1-2-38). If there is no guardrail on the operating platform, an operator may fall off. If there is no handrail on both sides of the inclined ladder used by the operator to climb up and down, the operator may fall and be injured.

FIG. 1-2-37 More than one operator turns the hand wheel at the same time (A).

FIG. 1-2-38 More than one operator turns the hand wheel at the same time (B).

FIG. 1-2-39 The operating platform's area is small, no guardrails and handrails (A).

FIG. 1-2-40 The operating platform's area is small, no guardrails and handrails (B).

FIG. 1-2-41 The operating platform's area is small, no inclined ladders.

FIG. 1-2-42 The inclined ladder has no handrails.

Remedy

In order for two operators to be able to stand on the operating platform at the same time, the platform must be at least 0.8 m × 1.2 m. There should be a guardrail not lower than 1 m on the operating platform. The ladder used by operators should be inclined rather than straight. The gradient of the inclined ladder should be 45° to 55°; the footstep should be horizontal, and 0.15 m to 0.3 m wide; and there should be a 1-m-high handrail on both sides of the inclined ladder (Figs. 1-2-39 to 1-2-44).

FIG. 1-2-43 Recommended operating platform (A).

FIG. 1-2-44 Recommended operating platform (B).

Hidden danger: The manual hand wheel is without an outer circle

Hazard

Turning the manual lever requires a large torque. When the hand wheel is without an outer circle or the diameter is too small, the torque is too small and turning is not easy. In Figure 1-2-45, the quadrilateral shape of the hand wheel and too small a radius lead to a small torque. In Figure 1-2-46 a wrench is used, making turning inconvenient, and creating a risk of fracturing the wrench from the root.

FIG. 1-2-45 Hand wheel radius is too small.

FIG. 1-2-46 A wrench is used.

Remedy

The hand wheel should have an outer circle with an outer diameter not less than 600 mm. To increase the opening force, and to make turning the hand wheel more convenient, four 10-cm-long handles can be arranged diagonally at each section of the outer circle (Figs. 1-2-47, 1-2-48).

FIG. 1-2-47 Recommended shape of hand wheels (A).

FIG. 1-2-48 Recommended shape of hand wheels (B).

Defect: The manual hand wheels are vertically arranged or the heights of the hand wheels from the operating platform are inconsistent

Hazard

The BOP manual hand wheels are vertically arranged, and the hand wheel heights from the operating platform are inconsistent. A single platform is not enough to meet the requirements of operating the three uneven hand wheels, leading to inconvenience of operating. For wells with three BOP stacks, to make the operation more convenient, some drilling teams have to set up two platforms (Figs. 1-2-49 to 1-2-52).

FIG. 1-2-49 Hand wheel heights vary more than 1.2 m.

FIG. 1-2-50 Erection of two platforms.

FIG. 1-2-51 Two vertically arranged hand wheels.

FIG. 1-2-52 Three vertically arranged hand wheels.

If the BOP manual hand wheels are vertically arranged and the hand wheels are in the same plane, there may be interference between each hand wheel, affecting the rotation, and it may even lead to the hand wheel not rotating (Fig. 1-2-53). To make the hand wheel turn flexible and avoid mutual interference between the hand wheels, staggering the hand wheels vertically by changing the lever lengths is usually introduced, as shown in Figure 1-2-54. With this layout, it is difficult for the platform to meet the needs of turning all the hand wheels.

FIG. 1-2-53 Hand wheels seem to be in the same plane.

FIG. 1-2-54 Hand wheels are staggered vertically.

Remedy

Without affecting the rotation of the manual lever, the lever can be deflected at an appropriate angle. The hand wheels should be arranged as horizontally as possible, reducing the height difference between them, making it easy for the operator to turn all the hand wheels on a single operating platform. Using one or two universal joints can make the horizontal arrangement of hand wheels easy. In the case of horizontal hand wheel layout, the length of the platform should be great enough so that an operator standing on the platform is able to turn all the hand wheels.

Almost all the Chinese well teams use a vertical hand wheel layout, whereas in Kazakhstan, a horizontal layout is common, which is more conducive to the operation of hand wheels (Figs. 1-2-55, 1-2-56).

FIG. 1-2-55 Three hand wheels in a horizontal layout.

FIG. 1-2-56 Two hand wheels at the same horizontal level.

Defects: The manual lever support is not solid enough at the hand wheel side; too large degree of freedom

Hazard

The manual operation lever support is not solid at the hand wheel side, and the degree of freedom is large. When turning the hand wheel, it swings up and down or side to side, which is not conducive to rotating (Figs. 1-2-57, 1-2-58).

FIG. 1-2-57 Manual lever support is not solid, easy to swing.

FIG. 1-2-58 Manual lever support is not strong, can be moved upward.

Remedy

The hand wheel side of the manual lever should be firmly supported and limited appropriately, making it easy to turn the hand wheel (Figs. 1-2-59, 1-2-60).

FIG. 1-2-59   Lever firmly supported (A).   **FIG. 1-2-60**   Lever firmly supported (B).

Hidden danger: The signboard on the manual lever hand wheel does not identify the specific switch circles (loops/circles/laps) number

Hazard

There are two major types of hidden dangers on signboards indicating the open or closed turns as follows:

1. Lack of specific open or closed turns. In Figures 1-2-61 and 1-2-62, signboards are marked "open or closed in place, rotate 1/4 to 1/2 circle backwards," "forward rotate in place, rotate 1/4 to 1/2 circle backwards," omitting specific values of open or closed. If the hand wheel "sticks" while turning, the crew may mistake it for a "place" (Figs. 1-2-61, 1-2-62).
2. The switch turns number given on the signboard is a range, not a specific value. Figures 1-2-63 and 1-2-64 are marked "opening 21 to 23 circles, turning off 21 to 23 circles." Is it fully turned off with 21 circles, or 23 circles? And the signboard is prefabricated—it is obviously not accurate to use the same value for any BOP open or closed turns. Different BOPs use different open or closed circles. Even with the same type of BOP from the same manufacturer, the switch turns may have some differences.

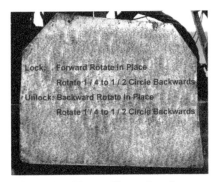

FIG. 1-2-61 The specific number of open or closed turns is not indicated (A).

FIG. 1-2-62 The specific number of open or closed turns is not indicated (B).

FIG. 1-2-63   Number of turns is a range (A).

FIG. 1-2-64   Number of turns is a range (B).

Remedy

To provide the operator with an accurate number of turns so that he will know whether the ram is open or closed, not in place, or is in the wrong open or closed direction, the signboard on the manual lever's hand wheel should indicate an open or closed direction, and the actual measured number of switching turns (Fig. 1-2-66). Open or closed turns to the end should be accurate to a quarter circle. And it is recommended that the type of gate is also labeled on the signboard (Fig. 1-2-65).

Defect: The hand wheel is without an automatic counting device

Hazard

At present most of the well rig crew teams rely on human memory to count the number of open or closed cycles when turning the hand wheel operator. This method often leads to errors (Fig. 1-2-67). Some other well teams tie counting ropes on the lever, and check the open or closed status by counting the number of rope loops. This approach does not provide real-time open or closed turns, and the counting method is cumbersome and error-prone (Fig. 1-2-68).

FIG. 1-2-65 Signboard indicating open or closed (A).

FIG. 1-2-66 Signboard indicating open or closed (B).

FIG. 1-2-67 Relying on human memory to count the number of open or closed turns.

FIG. 1-2-68 Counting with counting ropes.

BOP manual locking lever open or closed turns are generally drilled between 15.5 to 40 circles. An inaccurate open or closed turns count may lead to the lock shaft being left open, or not being closed in place. Not closing the locking shaft in place affects the sealing of the ram; not opening it in place may damage the ram.

Remedy

The BOP hand wheel mechanical counter should be used. It is an easy, fast, accurate, and reliable control opening and closing turns at any time, avoiding counting inaccuracies and errors.

The BOP mechanical counter is a device used on BOP manual locking to count the numbers of circles. Its principle is shown in Figure 1-2-69; it consists of a sensor that records the manual locking circle number (7), and the meter or indicator for the display of the circle number (1, 2, 3) and soft shaft (4). The number of hand wheel turning circles is passed from the sensors (a pair of gears) through transmission shafts to a secondary instrument header (a pair of worm wheel and worm), and through a pointer connected to the worm shaft. It shows the manual locking hand wheel rotation circles on the dial scale, and features a simple structure, counting accuracy, reliability, and easy installation, usage, and maintenance (Fig. 1-2-70).

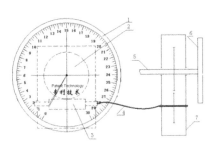

FIG. 1-2-69   Principle chart of a mechanical counter. 1—dial; 2—worm wheel; 3—worm; 4—soft shaft; 5—lock shaft; 6—lock wheel; 7—sensor.

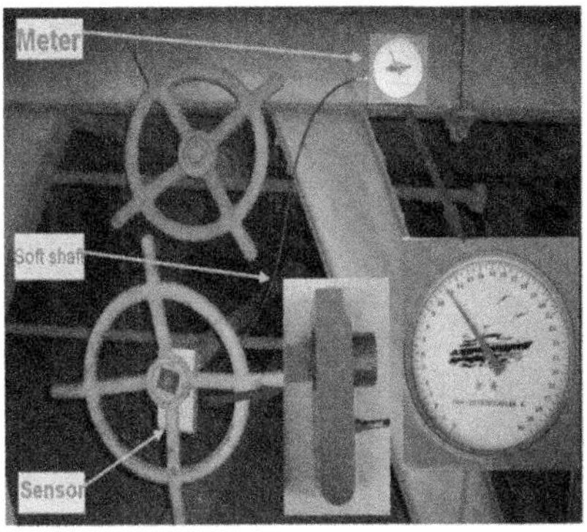

FIG. 1-2-70   Installing chart of a mechanical counter.

Hidden danger: Rotating a quarter- to a half-circle backward after being manually shut-in or locked in place

Hazard

When shutting-in ram BOPs with manual locking mechanisms, the usual practice is to lock and unlock to the end, then rotate one quarter- to one half-circle backward. For example, BOPs from Factory A, with a locking shaft clearance of generally 7 to 8 mm, rotating one quarter- to one half-circle on both sides after shutting-in, will leave a 4 to 8 mm gap between the rams. This gap is sufficient to prevent a BOP from sealing. On the manual for this BOP, the only requirement is to rotate backward after unlocking; it is not required to rotate backward after locking (Figs. 1-2-71, 1-2-72).

In many oil fields, when a manual ram BOP was closed under pressure, after rotating one quarter- to one half-circle backward, the ram BOP encountered sealing failures. Whether for a manual or hydraulic BOP, rotating backward after locking the ram is not conducive to sealing, and not necessary.

Remedy

After a ram BOP manual is shut-in or locked in place, it should not be rotated a quarter- to a half-circle backward. In order to facilitate a flexible start the next time, after unlocking counter clockwise in place, it can be rotated a quarter- to a half-circle backward.

FIG. 1-2-71   Rotating ¼- to ½-circle backward (A).

FIG. 1-2-72   Rotating ¼- to ½-circle backward (B).

3.   BOP STACK FIXATION

Hidden danger: Guy/anchor lines overhang the lower part of the BOP or around the body

Hazard

The higher the radial, the greater the displacement. It will be hard to keep the BOP from sloshing or to keep it steady if the anchor line is fastened to the lower part of the BOP or spool (Figs. 1-3-1, 1-3-2).

FIG. 1-3-1   Guy line position fastened too low (A).

FIG. 1-3-2   Guy line position fastened too low (B).

The anchor line is fastened to the BOP body at a lower position and off the diagonal line of the derrick substructure with poor stability, as shown in Figures 1-3-3 and 1-3-4.

FIG. 1-3-3 Guy lines rounded on the BOP body (A).

FIG. 1-3-4 Guy lines rounded on the BOP body (B).

Remedy

Fasten one end of the anchor line to the upper part of the BOP stack as far as possible, such as to the BOP flange, conductor flange, the flange between the annular BOP and the ram preventer, and the ear of the annular BOP. Do not fasten it to the cross joint or ram preventer (Figs. 1-3-5 to 1-3-8).

FIG. 1-3-5 Guy lines fastened to the BOP upper flange.

FIG. 1-3-6 Guy lines fastened between the annular BOP and ram preventer.

FIG. 1-3-7 Guy lines fastened to the bottom of the conductor.

FIG. 1-3-8 Guy lines fastened to the ears of the annular BOP.

Hidden danger: The guy line and BOP are not anchored

Hazard

The displacement and shaking of the BOP can't be prevented without anchoring. BOP stability is poor (Figs. 1-3-9, 1-3-10).

FIG. 1-3-9 Guy lines fastened to the ear of the BOP.

FIG. 1-3-10 Guy lines fastened to the ear of the annular BOP.

Remedy

The BOP stack should be anchored with guy lines to prevent it from shaking (Fig. 1-3-11). The method shown in Figure 1-3-12 is recommended. The BOP special clamp is used at the BOP flange, a method that is simple, convenient, fast, and low-cost.

FIG. 1-3-11 Guy lines anchored to the ear of the annular BOP.

FIG. 1-3-12 Guy lines anchored to the special collar clamp of the BOP.

Hidden danger: Fasten the anchor line without a steamboat ratchet or using a chain fall

Hazard

It is difficult to adjust an anchor line's tightness and centralize the BOP in real time without a steamboat ratchet, and the centralization and stability of the BOP cannot be ensured (Figs. 1-3-13, 1-3-14).

FIG. 1-3-13 Guy lines do not use a steamboat ratchet (A).

FIG. 1-3-14 Guy lines do not use a steamboat ratchet (B).

A chain fall shouldn't be used as a fastener. Although it is convenient for adjusting tightness, in time, a chain fall will fail to self-lock when applied with pull (Figs. 1-3-15, 1-3-16).

FIG. 1-3-15 Using a chain fall to guy lines (A).

FIG. 1-3-16 Using a chain fall to guy lines (B).

Remedy

By using a steamboat ratchet, the tightness of guy lines and centralization of the BOP can be adjusted at any time. It is inadvisable to replace a steamboat ratchet with a chain fall (Figs. 1-3-17, 1-3-18).

FIG. 1-3-17 Using a steamboat ratchet (A).

FIG. 1-3-18 Using a steamboat ratchet (B).

Hidden danger: The guy line diameter is less than 16 mm

Hazard

The smaller a guy line's diameter, the less stiffness and strength it provides. Plastic deformation will get more severe with poor fixed effect (Figs. 1-3-19, 1-3-20).

FIG. 1-3-19 Wire rope diameter is less than 16 mm (A).

FIG. 1-3-20 Wire rope diameter is less than 16 mm (B).

Remedy

The wire rope diameter should be at least 16 mm (Figs. 1-3-21, 1-3-22).

FIG. 1-3-21 Wire rope diameter is not less than 16 mm (A).

FIG. 1-3-22 Wire rope diameter is not less than 16 mm (B).

Hidden danger: Guy lines are not laid along a diagonal line of the derrick substructure

Hazard

It is difficult to restrain the BOP from shaking when guy lines are not laid along a diagonal line of the derrick substructure (Figs. 1-3-23 to 1-3-26).

FIG. 1-3-23 Guy lines are not laid along a diagonal line (A).

FIG. 1-3-24 Guy lines are not laid along a diagonal line (B).

FIG. 1-3-25 Guy lines intersect and are off a diagonal line (A).

FIG. 1-3-26 Guy lines intersect and are off a diagonal line (B).

The anchor line intersects and is off the diagonal line.

Remedy

The greatest stability occurs when guy lines are tightened on the four diagonal lines of the derrick substructure (Figs. 1-3-27, 1-3-28).

FIG. 1-3-27 Fasten on the diagonal line of the derrick substructure (A).

FIG. 1-3-28 Fasten on the diagonal line of the derrick substructure (B).

Hidden danger: Fasten guy lines (prop) to coping bolts of the annular preventer, side door bolts, or flange bolts

Hazard

One side of the guy lines (or prop) are anchored to the bolts of the BOP. Bolts are applied with a side force. Under side impact, nuts can easily become loose, or the load applied by the bolts decreases, which causes the bolts' tightness to be uneven, decreasing or losing sealing preferment. In time, bolts will bend and deform under side impact (Figs. 1-3-29 to 1-3-32).

FIG. 1-3-29 Guy lines fasten to the coping bolts of the annular preventer.

FIG. 1-3-30 Guy lines fasten to the side door.

FIG. 1-3-31   One end of the prop is fastened by bolts of the BOP.

FIG. 1-3-32   One end of the prop is fastened by bolts of the BOP.

Remedy

Fasten one end of the guy lines (props) to the BOP top flange, conduit bottom flange, flanges between the annular BOP and the ram preventer, and the ears of the annular BOP. Avoid applying the BOP fixed bolts with side force.

Hidden danger: Guy lines are fixed upward or horizontally

Hazard

When the guy lines are fixed above the BOP's foundation, it's not propitious for the seal between the BOP and spool, and the spool and casing head. Some people think guy lines should be fixed horizontally, but at a critical state, this is not feasible (Figs. 1-3-33, 1-3-34).

FIG. 1-3-33   The guy lines are fixed upward (A).

FIG. 1-3-34   The guy lines are fixed upward (B).

Remedy

According to mechanical principles, fastening one end of the guy lines to the upper part of the BOP stack and the another end to the lower part of the derrick substructure will provide the most stability, as shown in Figures 1-3-27 and 1-3-28.

Hidden danger: The guy lines knot up when used without rope clamps, when rope clamps are fastened reversely, or when the clamp interval is not standard

Hazard

The wire rope is knotted, U-clamps that are stuck on the main rope will damage the wire rope, and a clamp interval that is more or less than that of the standard will decrease the fixing strength (Figs. 1-3-35, 1-3-36).

FIG. 1-3-35 Using two rope clamps, clamps are fastened.

FIG. 1-3-36 The wire ropes tie a knot, without rope clamps.

Remedy 1

The clamp seat should be fastened on the servicing section, and the U-bolt should be fastened to the tail section. Rope clamps are not laid crossly. The interval of rope clamps should be about six to eight times the rope diameter (the clamp interval of 16 mm wire rope should be 96–128 mm). Another standard requires a clamp interval of 150 to 200 mm. The fastened extent should take wire rope deformation to one-third as standard. The length of the rope end should not be less than 140 mm from the last rope clamp, and it should be tied up with wire (Figs. 1-3-37, 1-3-38).

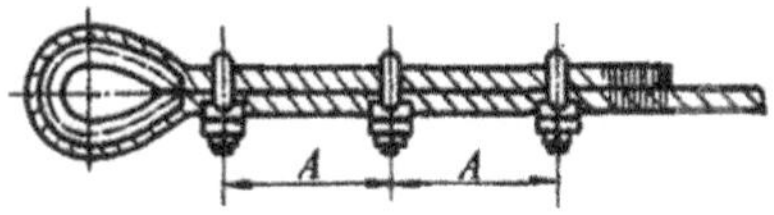

FIG. 1-3-37 Fastening methods of rope clamps.

FIG. 1-3-38 Normative rope clamps.

Remedy 2

Because of its convenience, safety, and no clamps, the pressed wire rope rigging has become more and more popular on site instead of wire rope (Figs. 1-3-39, 1-3-40).

FIG. 1-3-39 Pressed wire rope sling.

FIG. 1-3-40 Pressed wire rope rigging with steel jacket.

Choose the pressed wire ropes that meet the BOP's fastening requirements (Fig. 1-3-42). Wire rope rigging can't be used if it's too short and can't be used if it's beyond the range of the steamboat ratchet's adjustment. In Figure 1-3-41, the wire rope rigging is too long to be fastened to the earrings of the special flange and it is arranged on a diagonal line of the derrick substructure. So it has to be entwined around the flange.

FIG. 1-3-41 Pressed wire rope rigging is too long.

FIG. 1-3-42 Pressed wire rope rigging is moderate.

Defect: Fasten BOP stack by props

Hazard

During the second spudding, the Kelly will get incurved and run into the BOP at the rotary table's highest speed. The bad centralization and cement bond will intensify the BOP's shaking. Because the prop is inflexible, there will be a concentration of stress between the prop and BOP when it is shaking, which will cause a fracture, as shown in Figures 1-3-43 and 1-3-44.

FIG. 1-3-43 Fasten the BOP stack by props (A).

FIG. 1-3-44 Fasten the BOP stack by props (B).

Remedy

Using wire rope will fix this, because wire rope has some flexibility and can absorb shaking, alleviating stress concentration between the wire rope and BOP. The stress concentration can also be avoided by using a hinge (Figs. 1-3-45, 1-3-46).

FIG. 1-3-45 Connecting guy lines and BOP with hinges.

FIG. 1-3-46 Fasten the BOP by wire rope.

Defect: The wire rope winds around the BOP's attachment flange

Hazard

It takes longer to fasten when wire ropes wind around the BOP's attachment flanges, and it is inconvenient to check or fasten the BOP's binding bolts (Figs. 1-3-47, 1-3-48).

FIG. 1-3-47 Wire rope wound around the BOP's attachment flanges (A).

FIG. 1-3-48 Wire rope wound around the BOP's attachment flanges (B).

Remedy

Fasten a special flange to the joint of the BOP and conduit or a special collar clamp to the attachment flange of the ram preventer. Both ways are convenient and reliable (Figs. 1-3-49, 1-3-50).

FIG. 1-3-49 Special flange for BOP fastening.

FIG. 1-3-50 Special collar/clamp for BOP fastening.

4. MUD FENDER AND OTHERS

Hidden danger: The mud fender is horizontal or umbrella-shaped

Hazard

When the surface of a mud umbrella is horizontal or umbrella-shaped, drilling fluid from the derrick floor will fall onto the area around the well, contaminating the BOP and the spool. This creates unsanitary conditions under the platform and for equipment maintenance. Working under these conditions may cause injuries from slipping and falling (Figs. 1-4-1 to 1-4-4).

FIG. 1-4-1 Horizontal mud umbrella (A).

FIG. 1-4-2 Horizontal mud umbrella (B).

FIG. 1-4-3   Horizontal mud umbrella (C).   **FIG. 1-4-4**   Umbrella-shaped mud umbrella.

Remedy

To make collecting drilling fluid convenient, the mud blocking umbrella is designed in a reverse-umbrella shape; one or two openings at the bottom are connected with piping to discharge the drilling fluid that falls into the mud umbrella. The piping leads to the well or to other special fluid collectors to keep the area around the well and under the platform clean. For derricks that have a low mounting, which renders a reverse-umbrella-shaped fender impossible, a horizontal mud umbrella can be employed, but a circular block must be installed along the umbrella rim to hold drilling fluid (Figs. 1-4-5 to 1-4-8).

FIG. 1-4-5   Reversed mud umbrella (A).   **FIG. 1-4-6**   Reversed mud umbrella (B).

FIG. 1-4-8   Horizontal mud umbrella rounded

FIG. 1-4-7   Reversed mud umbrella (C).   with a block.

To prevent the mud umbrella from losing its balance, four corners of the mud fender can be fastened to the mounting by cable.

Hidden danger: The area of the mud umbrella is too small

Hazard

When the area of the mud umbrella is too small to collect all the drilling fluid from the platform, a portion of the fluid will fall on the BOP pack and valves at the well surface and under the platform, causing contamination (Figs. 1-4-9, 1-4-10).

Remedy

FIG. 1-4-9 The mud umbrella area is too small (A).

FIG. 1-4-10 The mud umbrella area is too small (B).

Enlarge the fender area until it is broad enough to collect all the drilling fluid from the platform, as shown in Figures 1-4-5, 1-4-7, and 1-4-8. If the fender area cannot be enlarged to the required extent, measures can be taken above the fender to direct the drilling fluid into the fender to keep the well surface and under the platform clean.

FIG. 1-4-11 Drilling fluid collection by guiding flow funnel.

FIG. 1-4-12 Drilling fluid collection by plastic cloth.

As shown in Figure 1-4-11, a guide filter is designed under the rotary table, which leads drilling fluid into the mud umbrella. Figure 1-4-12 shows fire-retardant plastic circles around the turntable, which lead drilling fluid into the fender.

Hidden danger: The BOP is wrapped with a plastic cloth instead of using a mud umbrella

Hazard

Wrapping the blowout preventer with plastic cloth can protect the BOP from being contaminated by drilling fluid, but the area under derrick floor is still vulnerable. Wrapping up the BOP makes daily inspection and maintenance inconvenient and potential dangers are not easily visible (Figs. 1-4-13, 1-4-14).

FIG. 1-4-13 Wrapping the BOP with plastic cloth (A).

FIG. 1-4-14 Wrapping the BOP with plastic cloth (B).

Remedy

Using a mud umbrella is the most effective way of keeping equipment at the well surface and under the platform clean. These areas should not be wrapped with plastic.

Hidden danger: The rectangular well (cellar) lacks an operation platform

Hazard

There are one or two hand-control valves beside not on the spool that require regular maintenance. When a hydraulic pipe leaks, repairmen are required. Valves should be closed when a blowout happens. If there is no operation plat-form on a rectangular well (cellar) (Figs. 1-4-15, 1-4-16), there is no room for staff to perform maintenance and repairs, such as turning the valves and repairing hydraulic lines, and there is even the danger of falling.

FIG. 1-4-15 A rectangular well with no operation platform (A).

FIG. 1-4-16 A rectangular well with no operation platform (B).

Remedy

To prevent falling and to make it easier to do repairs and maintenance, a operation platform should be set up above the rectangular (cellar) well, and the platform should be completely covered, without holes greater than 59 mm in diameter.

When the depth of a rectangular (cellar) well exceeds two meters, operations above the well are considered high-altitude operations. According to high-altitude operation standards, the cellar well and the platform should be solidly covered, without holes greater than 59 mm in diameter (Figs. 1-4-17, 1-4-18).

FIG. 1-4-17 Strictly covered operation platform for the cellar.

FIG. 1-4-18 Strictly covered operation platform for the rectangular well.

Chapter 2

Common Hidden Dangers and Remedies for Blowout Preventers (BOPs)

Common hidden dangers of hydraulic pipe installation:

- The straight sub is used to connect with the hydraulic lines and blowout preventer.
- The sealing of the hydraulic pipe sub loses efficacy.
- The hydraulic pipe is laid hidden in the earth.
- The hydraulic pipe is under the ground or compressed by vehicles.
- The distance between hydraulic pipe and relief pipe is less than 1 m.
- The distance of the hydraulic hose in the double-row rack is too close.
- The sub baffle of the double-row rack is fixed with bolts.
- There are no measures to prevent pollution by hydraulic oil at the hydraulic pipe sub.

Common hidden dangers of the remote console:

- The restricted-position device of the reversing valve is used in the pipe-ram BOP.
- The restricted-position device of the reversing valve is used in the blind-ram BOP.
- The restricted-position device of the reversing valve is used in the shear-ram BOP that has the function of a blind.
- The hydraulic pipe in the remote console leaks oil.
- The installation of the oil atomizer and water-gas separator is inclined.
- The remote console has either too much or not enough oil.
- The turning direction of the reversing valve of the remote console and BOP console and the open or closed direction are identical.
- Only the assistant driller is allowed to operate in the remote console.
- The access road of the remote console is less than 2 m wide.
- The distance between the remote console and wellhead is less than 25 m.
- Flammable or explosive goods (such as oil tanks, etc.) are within 10 m of the remote console.

- The remote console is placed in front of the derrick floor.
- The direction of the protection house doors of the remote console vary.

Common hidden dangers of the BOP console:

- The shear-ram BOP can be remote controlled on the BOP console.
- The reversing valve of the blind-ram BOP in the remote console is not installed with the accidental operation safety baffle.
- There is no break-proof device on the BOP/rig brake.

1. INSTALLATION OF HYDRAULIC PIPE

Defect: The hydraulic hose and BOP are connected with the straight sub

Hazard

The axial center line of the hydraulic pipe connector is horizontal. The hose and BOP are connected with the straight sub, and the hose is bent by deadweight. Stress concentration is produced at the bent position of the hose. Long-term stress may cause the hose to crack until it is damaged (Figs. 2-1-1, 2-1-2).

FIG. 2-1-1 Hydraulic hose bent and broken.   **FIG. 2-1-2** Hydraulic hose bent.

Remedy 1

If the hydraulic hose and BOP are connected with a bent sub at a 90° right angle, the stress concentration caused by the bent hose can be avoided. It is important to protect the hydraulic hose. The bent extent of the hose is related to the bent angle of the servicing sub. The bent extent of the hose is greatest if the straight sub is applied. If the 120° bent sub is used, the hose will still be bent, as in

Figure 2-1-4. When the bent angle is 90°, the hose is not bent, so a 90° bent sub is the best choice for protecting the hose, as shown in Figure 2-1-3.

A 90° bent sub that turns 360° is installed simply and conveniently.

Remedy 2

The closed bent sub or tee bent sub is used at the BOP connector/interface, and it is also a turning manner (Figs. 2-1-5, 2-1-6). The closed bent sub or tee bent sub is rigid pipe, and although the connection is safe and reliable, pipe is not as good as hose in distribution and installation. The closed bent sub or tee bent sub will be replaced gradually by hydraulic hose, which installs simply and conveniently and is reliably safe.

FIG. 2-1-3   90° bent sub is used. 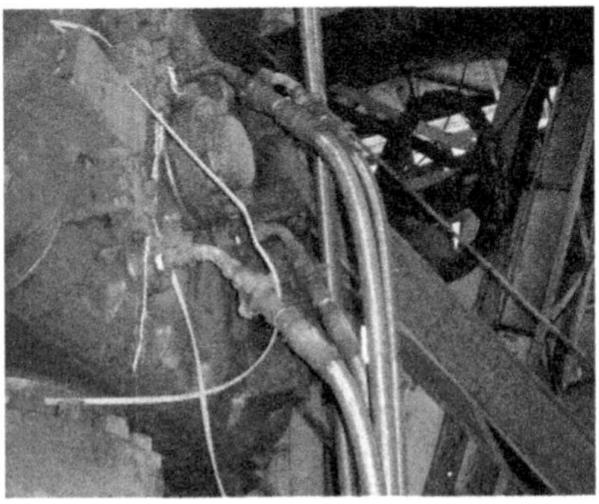   **FIG. 2-1-4**   120° bent sub is used.

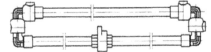   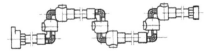

FIG. 2-1-5   Closed bent sub.

FIG. 2-1-6   Tee bent sub.

Hidden danger: Failure to seal the hydraulic hose sub

Hazard

The union is not sealed tightly or the self-sealing union is damaged. Hydraulic oil will leak from the sub. At the drilling site, the failure to seal the hydraulic hose sub occurs often. Currently, the methods of reversing valve handles and opening or closing control targets are identical.

The hydraulic line is under a pressure of 10.5 MPa for a long time. The failure to seal the hydraulic hose sub is more universal. Not only does an oil leak affect the control system, but it also pollutes the environment (Figs. 2-1-7, 2-1-8).

FIG. 2-1-7 The sub of a hydraulic pipe trickling.   **FIG. 2-1-8** The sub of a hydraulic pipe stabbing.

Remedy

In order to insure the sealing effect of the hydraulic hose sub, the union should be sealed tightly, and a seal filler with high-quality reliability should be used. After connecting, the 21 MPa test must be done because some hose subs still leak oil after being connected.

Hidden danger: The hydraulic line is buried, on the ground directly, or compressed by vehicles

Hazard

Hydraulic hose is buried underground, which makes it difficult to discover oil leaks. Buried pipe underground is easily eroded (Fig. 2-1-9) and is inconvenient to maintain and inspect daily. If the region of buried pipe is compressed by vehicles, or something heavy is put on the surface of the buried pipe, it is likely to be damaged (Fig. 2-1-11).

When drilling in the rainy season, or in a rainy region, if hydraulic hose is placed on the ground directly, it will be buried and eroded by muddy water, which will shorten the service life of the pipe. It is also difficult to inspect and maintain (Figs. 2-1-10, 2-1-12).

FIG. 2-1-9 Hydraulic pipe is buried.   **FIG. 2-1-10** Hydraulic pipe is on the ground.

FIG. 2-1-11 Pipe is buried and compressed by vehicles.

FIG. 2-1-12 Muddy well site after rain.

Remedy

Hydraulic hose should be not be on the ground, especially when drilling in the rainy season or in a rainy region; it should be on the rack 10 cm above the surface, as shown in Figure 2-1-13.

Figure 2-1-14 shows some ways of laying hydraulic hose in the west, where rain is rare. Sandy soil is piled up about 10 cm from the earth's dais. Put a plastic cloth on the surface of the dais, and place the hydraulic hoses on the cloth, keeping the pipe out of muddy water. There must be protection measures in place, such as a passing bridge, so that vehicles can pass by, as shown in Figures 2-1-15 and 2-1-16.

FIG. 2-1-13 Hydraulic hose on the rack.

FIG. 2-1-14 Hydraulic hose on plastic cloth.

FIG. 2-1-15 Piping rack and passing bridge (A).

FIG. 2-1-16 Piping rack and passing bridge (B).

Hidden danger: The distance between the hydraulic piping and also flow line is less than 1 m

Hazard

Tools are needed when the hydraulic hose and relief line are connected or fixed. If the distance between the hydraulic hose and relief line is too close it will be too inconvenient for operators because they will not have enough operating room/space, as Figure 2-1-17 shows.

Remedy

The distance between the hose rack and the relief line should be greater than 1 m, as shown in Figure 2-1-18.

FIG. 2-1-17 The distance between the hydraulic hose and relief line is less than 1 m.

FIG. 2-1-18 The distance between the hose rack and relief line is greater than 1 m.

Defect: The distance between pipes of the double-deck pipe row rack is too close

Hazard

Figures 2-1-19 and 2-1-20 show a double-deck pipe row rack laid up and down vertically, at a distance of 9 to 11 cm. When the union sealing of the hose in the lower row is invalid, it needs to be hammered tightly. Because the oil hose is blocked by the upper deck, it is very difficult to hammer the union directly. Some drilling crews remove the unions of the upper deck hose row rack and hammer the leaking unions of the lower deck to tighten them, and then tighten the unions in the upper deck again.

Remedy

In order to connect piping conveniently, the single-deck pipe row rack or hydraulic hose should be used (Figs. 2-1-21, 2-1-22). If the double-deck pipe row rack is used, when the pipe row rack is designed and manufactured, the distance between oil pipes should be enlarged slightly so that it will be convenient to tighten or remove.

 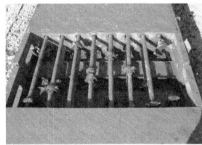

FIG. 2-1-19   Double-deck pipe row rack (A).   **FIG. 2-1-20**   Double-deck pipe row rack (B).

FIG. 2-1-21   Single-deck pipe row rack.   **FIG. 2-1-22**   Hydraulic hose.

Defect: The baffle of the pipe row rack is fixed with bolts

Hazard

After some drilling crews install the control device, they put the baffle above the union of the oil pipe, and secure it with bolts at four corners (some supervisors or managers ask for this too). It is difficult to inspect daily if the pipe-row rack baffle is fixed with bolts. In order to inspect the sealing of the union, the bolts must be screwed off first, then the baffle is opened, and the union sealing is inspected. If everything is in good condition, the baffle is recovered, the bolts are tightened again, and it returns to its original state. This inspection needs to be repeated many times each tour/shift. This method wastes time and is inconvenient (Figs. 2-1-23, 2-1-24).

FIG. 2-1-23 Baffle of pipe-row rack fixed with bolts (A).

FIG. 2-1-24 Baffle of pipe-row rack fixed with bolts (B).

Remedy

The function of the baffle is to protect the union of the oil hose so that it doesn't get damaged during transportation. It is unnecessary to fix the baffle with bolts again after the hose is connected. In order to prevent other objects from striking the union, and to prevent the hydraulic oil from injuring people, the baffle can be blocked with one bolt, as shown in Figure 2-1-25. The baffle can turn free, so it is convenient to inspect the pipe sealing. Figure 2-1-26 shows the baffle put on the pipe-row rack directly. Two handles are fitted on the baffle, which can be opened to inspect the pipe sealing when necessary.

FIG. 2-1-25 Only one bolt in the baffle is turning free.

FIG. 2-1-26 Baffle on the union.

Defect: There is no measure to prevent oil pollution at hydraulic unions

Hazard

It is hard to avoid pipe leakage during operation. Leakage will cause some hydraulic oil to spill on the ground. If there is no prevention measure at the sub, the hydraulic oil will pollute the environment (Figs. 2-1-27, 2-1-28).

FIG. 2-1-27 Ground contamination with hydraulic oil.

FIG. 2-1-28 No prevention measures for hydraulic oil falling to the ground.

Remedy

In order to avoid hydraulic oil polluting the environment, set up the oil trough or oil support and put seepage-proof plastic cloth under the pipe sub at the interface where the hydraulic pipe is connected or dismantled often in the remote console (Figs. 2-1-29 to 2-1-32).

FIG. 2-1-29 Set up oil trough.

FIG. 2-1-30 Set up oil support.

FIG. 2-1-31 Put seepage-proof plastic cloth under the pipe sub (A).

FIG. 2-1-32 Put seepage-proof plastic cloth under the pipe sub (B).

2.   REMOTE CONSOLE

Hidden danger: The position-limited device is used in the reversing valve of the pipe-ram BOP

Hazard

If the position-restricted device is used in the reversing valve of the pipe-ram BOP, the driller is unable to control it remotely (Figs. 2-2-1, 2-2-2).

FIG. 2-2-1   Location of the reversing valve of the pipe-ram is restricted (A).

FIG. 2-2-2   Position of the reversing valve of the pipe-ram is restricted (B).

Remedy

Neither uses the protector in the reversing valve of pipe-ram BOP, nor the position-restricted device (Fig. 2-2-3).

FIG. 2-2-3   No position-restricted device of the reversing valve of the pipe-ram BOP.

Hidden danger: The position-restricted device is used in the blind-ram BOP

Hazard

If the position-restricted device is used in blind-ram BOP, the driller is unable to control it remotely (Figs. 2-2-4 to 2-2-7).

FIG. 2-2-4   Position of the reversing valve of the blind-ram is restricted with a pin.

FIG. 2-2-5   Position of the reversing valve of the blind-ram is restricted with a baffle (A).

FIG. 2-2-6   Position of the reversing valve of the blind-ram is restricted with a baffle (B).

FIG. 2-2-7   Position of the reversing valve of the blind-ram is restricted with wire tying.

Remedy

The blind-ram BOP should be equipped with the safety cover that does not hold back remote control. During drilling operation, only the driller knows whether there is drill string in the borehole. Either close the pipe-ram BOP or the blind-ram BOP. The purpose of installing the safety cover in the reversing valve of the blind-ram BOP is to prevent any mistaken operation by the assistant driller in the remote console, instead of the driller's operation by mistake (Figs. 2-2-8, 2-2-9).

FIG. 2-2-8 Safety cover is used in the reversing valve of the blind-ram BOP (A).

FIG. 2-2-9 Safety cover is used in the reversing valve of the blind-ram BOP (B).

Defect: The position-restricted device is used in the reversing valve of the blind-shear-ram BOP

Hazard

In the BOP stack, if there is no blind-ram BOP, but only blind-shear-ram BOP, the reversing valve of the blind-shear-ram BOP should be equipped with the position-restricted device as shear-ram BOP. When the well is vacant, the driller is unable to carry out remote control, and it is unfavorable to shut-in on time and rapidly (Figs. 2-2-10, 2-2-11).

Remedy

FIG. 2-2-10 Baffle is used in the restricted position of the reversing valve (A).

FIG. 2-2-11 Baffle is used in the restricted position of the reversing valve (B).

Blind-shear-ram BOP can cut off the drill string and shut-in the borehole. When there is no drill string in the hole, it can function as a blind-ram BOP. When drilling design is done, the BOP stack is often simplified. Because the height of the substructure is restricted, a pair of blind-shear-ram BOPs are used to replaced the two blind-ram and shear-ram BOPs. Thus, when a pair of rams are used, it can ensure that the wellhead is sealed completely during vacant borehole operation, and it can ensure that the drill pipe is cut off and that it seals the wellhead under accidental circumstances. During normal drilling, blind-shear-ram BOP is used only as a blind seal function. It serves as blind-shear-ram BOP only in accidental circumstances.

If there is one pair of blind-shear-ram BOPs and no other blind-ram BOP is in the BOP stack, the blind-shear-ram can be regarded as a blind-ram BOP. The reversing valve should be handled as a blind-ram BOP, namely, the reversing valve should be equipped with the safety cover that does not hold back remote control. The position-restricted device is not used, having the function of remote control, as shown in Figures 2-2-12 and 2-2-13.

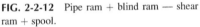

FIG. 2-2-12   Pipe ram + blind ram — shear ram + spool.

FIG. 2-2-13   Safety cover used in the reversing valve.

SY/T 5053.2—2001 regulates "3 ways 4 positions valve used in controlling blind-ram BOP (or shear-ram BOP) should be equipped with the safety cover that do not hold back remote control." *SY/T 6616—2005* regulates "three-location, four-passage reversing valve used in controlling the switch of shear-ram BOP in the BOP control equipment should be equipped with mistake-operation safety device. The controlling valve of operating shear-ram BOP should be not installed on BOP driller's console." The first standard allows remote operation, but the second standard does not. The requirements of the two standards for the same problem are not identical.

If the height of the derrick substructure can meet the BOP installation, it is inadvisable that the assembly of blind-ram BOP and shear-ram BOP is replaced by a pair of blind-shear-ram BOPs. In a western well, after the drill pipe was cut off, the wellhead could not be shut-in. It was risky to change the pipe-ram BOP into a blind-ram BOP, so the wellhead could be controlled, as in Figure 2-2-14. Be aware that when the blind-ram BOP was tested in the factory, it could cut off a drill pipe, and shut-in a wellhead. But on the drilling site, it is still very challenging to product quality that a shear-ram BOP could cut an approximately 100 t drill string. It does not deform or get damaged, and it has the ability to seal the wellhead.

FIG. 2-2-14   After cutting off the drill pipe, the shear-ram BOP can shut-in the wellhead.

Hidden danger: Hydraulic pipe in the remote console leaks oil

Hazard

There is oil leakage from equipment in the remote console. This will interfere with the normal function of equipment.

Remedy

No equipment in the remote console should be allowed to leak oil. Inspect at least once each tour/shift. If you discover an oil leak, repair it at once.

Hidden danger: The oil atomizer or water-gas separator in the remote console is installed on an incline

Hazard

There is not enough oil in the oil atomizer. The piston of the gas-drive hydraulic pump lacks lubrication, and if the pump services for a long time it will be damaged. Because the oil atomizer is inclined, it does not form oil drops easily, and the droplet effect is poor (Fig. 2-2-15). If the water-gas separator is inclined, the water filtration efficiency will be decreased (Fig. 2-2-16).

The principle of the oil atomizer is as follows: Oil flows from the nozzle by way of a high-speed air current that flows over the cylinder, causing the surface pressure of the oil drop to distribute unevenly. This pressure distribution makes the oil drop elongate until it is broken or changed into a lot of oil droplets, which are then broken into even smaller oil droplets. This continues until the pressure distribution on the oil droplet surface is less than the surface tension that tightens the surface of the oil droplet; therefore, the oil droplet cannot atomize further.

The principle of a water-gas separator is as follows: When compressed air flows in from the inlet of water-gas separator, liquid state oil, water, and impurities contained in the gas rotate violently along the liner direction of the impeller. Liquid state oil, water, and solid impurities are thrown to the inside wall of the water-holding cup under centrifugal force and flow to the bottom. The compressed air removes the liquid state oil, water, and impurities, and micro solid particles are cleaned out further through a filtration core and then flow out from the outlet. A water baffle is used to prevent liquid state oil and water accumulated from interfusing into the air current again. After operating the handle of the draining valve and lifting the draining plug, the cooling water is drained out through the annular space between the draining plug and sealing filler and the center hole.

FIG. 2-2-15 Oil atomizer and the water-gas separator is inclined (A).

FIG. 2-2-16 Oil atomizer and the water-gas separator is inclined (B).

Remedy

The oil atomizer and water-gas separator should be installed vertically, and have all the necessary and perfect fittings and normal functions. The oil atomizer drops oil at a rate of 5 to 8 drops per minute.

Hidden danger: There is too much or not enough oil in the remote console

Hazard

The minimum allowable oil level in the remote console oil tank is the minimum amount of oil to keep the remote console normally running. The amount of oil used in the accumulator group is approximately half of the effective volume. If the oil level in the tank exceeds the allowable maximum oil level, hydraulic oil will overflow from the oil-pouring port located on the top of the oil tank when the accumulator releases pressure (Figs. 2-2-17, 2-2-18).

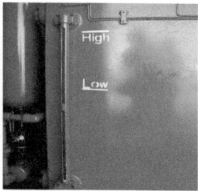

FIG. 2-2-17 Oil quantity is less than the allowable minimum level.

FIG. 2-2-18 Oil quantity is more than the allowable maximum level.

Remedy

Hydraulic oil in the oil tank of the remote console should be between the allowable minimum and maximum oil levels recommended by the factory (Figs. 2-2-19, 2-2-20).

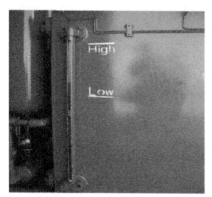

FIG. 2-2-19 Correct oil level (A).

FIG. 2-2-20 Correct oil level (B).

Defect: The turning direction of the control valves of the remote console and the driller's BOP control console comply with the working condition of the BOP

Hazard

According to industry standards and factory instructions for remote consoles, three-location, four-passage reversing valves of the remote console should be identical with the BOP open or closed during awaiting orders. That is, reversing valves should be open while awaiting orders. But the crew on site always hope that the reversing valves are in the middle location while they are awaiting orders. Operating staff and managers are often confused by why reversing valves are in the middle or open location.

The reversing valves are in the open location when they are awaiting orders. Closed pipe does not bear pressure, but open pipe bears 10.5 MPa pressure over time, raising the possibility of oil leaks at the sub. Once there is an oil leak, a lot of hydraulic oil will be lost. The switches of the electric pump in the remote console are usually in the pressure charged state. When the pressure of the remote console is less than 19 MPa, the system is supplemented with pressure automatically. The remote console is always in normal awaiting orders. If the hydraulic pipe leaks oil and it is not discovered, the electric pump will run until the oil in the tank gas is gone.

The reversing valves are in the open location when they are awaiting orders. The sealing circumstances of the open pipe sub only can be inspected; the sealing circumstances of the closed pipe sub can be inspected only when shutting-in. The sealing effect of the closed pipe sub influences whether the BOP can be closed in time and effectively (Figs. 2-2-21, 2-2-22).

FIG. 2-2-21   Reversing valve of the pipe-ram BOP in the open location (A).

FIG. 2-2-22   Reversing valve of the pipe-ram BOP in the open location (B).

The reversing valves are in the middle location when they are awaiting orders. Open and closed pipe do not bear 10.5 MPa pressure constantly. The possibility of leaking oil is very rare; even if leakage occurs, not a lot of hydraulic oil will be lost, and the function of the remote console will not be influenced. Of course, while reversing valves are in the middle location during awaiting orders, it is impossible to inspect the sealing circumstances of the closed pipe sub. The sealing circumstances of the closed pipe sub can be inspected only when shutting-in. So, when awaiting orders, reversing valves are not only in an open or middle location, but the closed pipe does not bear pressure constantly, and there is no influence to shut-in (Figs. 2-2-23, 2-2-24).

FIG. 2-2-23 Reversing valve of the blind-ram BOP in the middle location.

FIG. 2-2-24 Reversing valve of the pipe-ram BOP in the middle location.

The people who insist on reversing valves in the open location think that it is favorable to prevent the ram from stretching to the borehole when they are in the open location. Reversing valves are in the middle location during awaiting orders. When the ram closed pipe and accumulator open onto each other, oil flows back in the open pipe or the pipe leaks oil at the same time, and the ram may stretch to the center of borehole. Otherwise, it is unlikely to occur as described earlier. The phenomenon where the ram stretches to the center of the borehole automatically is seen rarely. The reason why the ram stretches to the center of the borehole is still disputed, so there is no actual conclusion when the locations of reversing valves are discussed.

The awaiting orders operating mode of the BOP reversing valves was discussed before. Next, the awaiting orders operating mode position of the reversing valves of the hydraulic relief valve is discussed. Because the standard only asks, "Turning directions of reversing valves of remote console, BOP driller's console and BOP open or closed are identical," the turning direction of the hydraulic relief valve is not regulated. Combined with the previous analysis, having the hydraulic relief valve in the middle location is favorable for protecting equipment, and also affects the function of equipment. Of course, this contradicts "the turning direction of the hydraulic relief valve and the open or closed state of the valve are identical" recommended by the factory (Figs. 2-2-25, 2-2-26).

The standard states that "the turning direction of reversing valve in BOP driller's console and the open or closed states of BOP are identical," which leads to a lack of maneuverability. The middle location function of the valve core of all reversing valves (gas turning valves) in the BOP driller's console are all Y model now, and it can recover automatically. All reversing valves are in the middle location while awaiting orders. So the reversing valve in the BOP console is not identical with the open or closed states of the BOP (Figs. 2-2-27, 2-2-28).

FIG. 2-2-25 Reversing valve in the middle location.

FIG. 2-2-26 Reversing valve in the closed location.

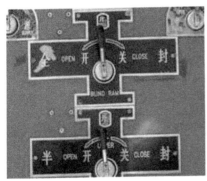

FIG. 2-2-27 Reversing valve of the driller's console in the middle location (A).

FIG. 2-2-28 Reversing valve of the driller's console in the middle location (B).

Remedy

Because of structure and function, the awaiting orders operating mode of the reversing valve in the BOP driller's console is only in the middle location. For the awaiting orders operating mode of reversing valves in the remote console used in controlling ram BOP, the hydraulic relief valve should be in the middle location.

In order to prevent pipe from leaking suddenly, currently, the awaiting orders operating mode of the reversing valves is still in the middle location on the drilling site. The valves are in the open location only when the inspector comes to the well site. After the inspector leaves, the valves are returned to the middle location. This action is not approved (Figs. 2-2-29, 2-2-30).

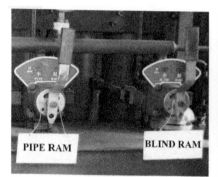

FIG. 2-2-29   Reversing valve in the middle location (A).

FIG. 2-2-30   Reversing valve in the middle location (B).

Defect: Only the assistant driller is allowed to operate in the remote console

Hazard

Currently, some oil fields stipulate that only the assistant driller is allowed to operate in the remote console. In normal conditions, however, it can be done (Figs. 2-2-31, 2-2-32). When neither the driller or assistant driller is on duty, the derrick man assumes the assistant driller position but is not allowed to operate the remote console. If, for any reason, the assistant driller cannot reach the remote console in time after an alarm, and other people are not allowed to operate the remote console, it is likely to cause an out-of-control blowout.

FIG 2-2-31   Only one person is allowed to operate the valve.

FIG. 2-2-32   Particular operation post is not appointed.

Remedy

The personnel in the remote console should have the same post responsibility as the assistant driller. Personnel in higher positions should also have a right to operate the remote console, to ensure that a wellhead can be controlled rapidly and on time under any special conditions.

3. ARRANGEMENT OF REMOTE CONSOLE

Hidden danger: The open road's width around the remote console is less than 2 m

Hazard

The width of the open road around the remote console is less than 2 m, which makes it inconvenient to maintain and repair the remote console (Figs. 2-3-1, 2-3-2).

FIG. 2-3-1   No passageway around the remote console.

FIG. 2-3-2   Blocked passageway around the remote console.

Remedy

The open road around the remote console should be spacious, greater than 2 m (Figs. 2-3-3, 2-3-4).

FIG. 2-3-3   The width of the remote console is more than 2 m (A).

FIG. 2-3-4   The width of the remote console is more than 2 m (B).

Hidden danger: The remote console is less than 25 m from the wellhead

Hazard

The remote console is too close to the wellhead. Once a blowout catches fire, it is difficult for operation personnel to approach the remote console because of the radiation of flame (Fig. 2-3-5). There is no way to shut-in, causing an out-of-control blowout and causing the derrick to fall, as Figure 2-3-6 shows.

FIG. 2-3-5 Remote console is about 10 m from the wellhead.

FIG. 2-3-6 Blowout is out of control, and the derrick has fallen down.

Remedy 1

The remote console is located 25 m beyond the wellhead (Figs. 2-3-7, 2-3-8).

FIG. 2-3-7 Remote console is located 25 m from the wellhead (A).

FIG. 2-3-8 Remote console is located 25 m from the wellhead (B).

Remedy 2

If the BOP auxiliary equipment is equipped at the entry of the well site door or in the cadre duty house, it is possible that the remote console is closer to the wellhead. In some oil fields in southeastern China, the well site environment is complex and sensitive, and the lease land is difficult. The well site area is smaller, so it is difficult that the remote console is 25 m from the wellhead. For these types of wells, the auxiliary controlling equipment can be equipped. The equipment should be arranged near the cadre's duty house where someone is on duty, or at the door of the well site. Once a blowout happens, the auxiliary equipment can be used to shut-in. An acceptable distance between the remote console and wellhead is 20 to 25 m.

Because some drilling crews abroad are equipped with BOP auxiliary controlling equipment, the remote console is closer to the derrick substructure (Fig. 2-3-10). In China, drilling crews are not equipped with BOP auxiliary controlling equipment except for marine/off-shore drilling and extreme individual exploratory deep wells. We suggest that wells containing high concentrations of hydrogen sulfide, super deep wells, and wells where it is very difficult for the well site area to meet the safety distance for the remote console be equipped with BOP auxiliary controlling equipment. In Figure 2-3-9, the remote console is closer to the substructure. Once the wellhead catches fire or the derrick falls, the remote console will be damaged. Even though there is auxiliary equipment, shut-in operation cannot be carried out.

FIG. 2-3-9 Remote console is near the sub-structure.

FIG. 2-3-10   Auxiliary equipment of the remote console.

Hidden danger: Flammable or explosive objects like oil tanks are within 10 m of the remote console

Hazard

Oil tanks and acetylene bottles are all flammable and/or explosive. If objects like this are within 10 m of the remote console, once they catch fire or explode, they will likely ignite the remote console, endangering the remote console safety (Figs. 2-3-11, 2-3-12).

FIG. 2-3-11   Oil tank is 2 m from the remote   **FIG. 2-3-12**   Acetylene house is about 1 m from
console.   the remote console.

Remedy

Nothing flammable or explosive should be within 10 m of the remote console,
as shown in Figure 2-3-3.

Hidden danger: The remote console is located in front of the derrick floor

Hazard

The remote console is located in front of the derrick floor and hydraulic pipe
goes through the pipe bridge or drill pipe rack. First, it is difficult to inspect
and replace hydraulic pipe, as Figures 2-3-13 and 2-3-14 show; second, there
is not enough room for forklift truck operation because of the location of the
hydraulic pipe. The drill pipe rack located on the left of the well site cannot
be used. This is rather inconvenient to a drilling operation, as shown in
Figures 2-3-15 and 2-3-16. Finally, during large operations like cementing,
etc., the location of the remote console in front of the derrick floor is inconvenient to the drilling operation.

Remedy

The remote console should be located on the left of the derrick floor (on the
left front or on the left rear) 25 m beyond the wellhead. It should not influence
the drill pipe rack on the left being used, as shown in Figures 2-3-17 and 2-3-18.
For an electrical drilling rig, the remote console can be located to the rear
of the derrick floor, 25 m from the wellhead, as shown in Figures 2-3-19
and 2-3-20.

FIG. 2-3-13 Remote console is located in front of the derrick floor (A).

FIG. 2-3-14 Remote console is located in front of the derrick floor (B).

FIG. 2-3-15 Remote console is located in front of the derrick floor (C).

FIG. 2-3-16 Remote console is located in front of the derrick floor (D).

FIG. 2-3-17 Remote console is located at the rear of the derrick floor.

FIG. 2-3-18 Remote console is located at the front of the derrick floor.

FIG. 2-3-19 Remote console of the electric rig is located at the rear of the derrick floor (A).

FIG. 2-3-20 Remote console of the electric rig is located at the rear of the derrick floor (B).

Defect: The door orientations of the remote console protection house are not identical

Hazard

The direction of the protection house door of the remote console is not regulated by the standard, nor by detailed rules and regulations of many oil fields. One point of view is that the protection house door of the remote console should face the derrick floor. That way the assistant driller can turn about and transfer sign signals to the derrick floor conveniently. Another point of view is that the protection house door should be at the back of the derrick floor. Once a well blows out or catches fire, the protection house can block part of the radiant heat, and it's easier for operating personnel to approach the remote console. Still another point of view is that the side of the remote console should face the derrick floor. It is convenient for the assistant driller to have contact with the roughneck on the derrick floor (Figs. 2-3-21 to 2-3-24).

FIG. 2-3-21 Protection house door is on the left front facing the floor.

FIG. 2-3-22 Protection house door is on the left rear behind the floor.

FIG. 2-3-23 Protection house door of the V-door is behind the floor.

FIG. 2-3-24 The side of the protection house door on the left rear facing the floor.

Remedy

On the same oil field, the arrangement of equipment should be standardized. The door orientation of the remote console should be identical. When arranging the remote console, first ensure a safe distance from the wellhead; second, ensure that no inflammable or explosive goods are within 10 m of the remote console. Last, the protection house door orientation of the remote console should be at the back of the derrick floor. Once a blowout happens and catches fire, the protection house can block part of the radiant heat. Operating personnel can approach the remote console conveniently to control the wellhead, as shown in Figure 2-3-22.

4. BOP DRILLER'S CONSOLE

Hidden danger: The shear-ram BOP can be controlled remotely on the driller's console

Hazard

When a blowout is out of control, the drill string needs to be cut off, and the shear-ram should be used. It is very dangerous to control the shear-ram BOP on the driller's console remotely. When the drill string is cut off instantly, the hoisting system loses weight suddenly and makes the derrick and equipment vibrate intensely, causing objects to fall, which can cause injuries (Figs. 2-4-1, 2-4-2).

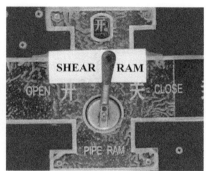

FIG. 2-4-1 Shear ram is controlled in the driller's console (A).

FIG. 2-4-2 Shear ram is controlled in the driller's console (B).

Remedy 1

The driller cannot remotely control the shear-ram BOP; it can be operated only in the remote console. The location-restricted device should be used on the reversing valve of the shear-ram BOP in the remote console (Figs. 2-4-3 to 2-4-6).

FIG. 2-4-3   Location used in reversing valve (A).

FIG. 2-4-4   Location used in reversing valve (B).

FIG. 2-4-5   Location used in reversing valve (C).

FIG. 2-4-6   Location used in reversing valve (D).

Remedy 2

If a blind-ram BOP is not disposed in the BOP stack, and a blind-shear-ram BOP is equipped in the stack, the driller can be allowed to remotely control it in the driller's console. That is, in a state of well naught, the driller can shut-in from the driller's console, but when the drill string needs to be cut off, the staff should leave the derrick floor, and operate from the remote console.

Hidden danger: The reversing valve of the blind-ram BOP is not installed with a safety baffle that prevents accidental operation

Hazard

Because the reversing valve of the blind-ram BOP is not installed with a safety baffle that prevents accidental operation, it is likely to mistakenly operate the blind-ram BOP (Figs. 2-4-7, 2-4-8).

FIG. 2-4-7 Reversing valve is not installed with a safety baffle (A).

FIG. 2-4-8 Reversing valve is not installed with a safety baffle (B).

Remedy

In order to prevent a driller from mistakenly operating the blind-ram BOP on the driller's console, the safety baffle should be installed in the reversing valve of the blind-ram BOP in the driller's console (Figs. 2-4-9 to 2-4-12).

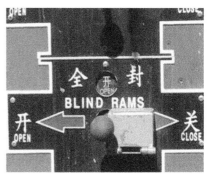

FIG. 2-4-9 Safety baffle is installed in the reversing valve of the blind-ram BOP (A).

FIG. 2-4-10 Safety baffle is installed in the reversing valve of the blind-ram BOP (B).

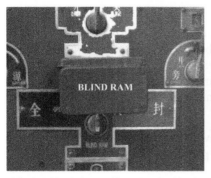
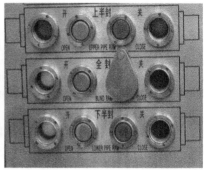

FIG. 2-4-11 Safety baffle is installed in the reversing valve of the blind-ram BOP (C).

FIG. 2-4-12 Safety baffle is installed in the button reversing valve of the blind-ram BOP.

Hidden danger: There is no BOP/gear break-proof safety device for the rig brake (drill rig break-proof device)

Hazard

A BOP/gear break-proof safety device of a rig brake (drill rig break-proof device) is not installed. Once operating accidentally, the drill string is raised in the case of a closing pipe-ram BOP, the wellhead will be pulled out, the drill pipe and drill collar will break, and the derrick is pulled down. Once this occurs, loss will be enormous. Even today, this accident often occurs. Figure 2-4-13 was an accident that occurred abroad a while ago. The assistant driller accidentally closed the pipe-ram BOP, which pulled down the derrick.

FIG. 2-4-13 Pulling down the derrick from accidental operation.

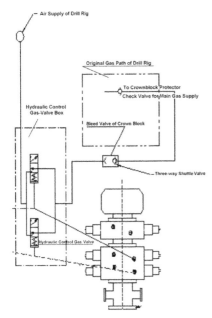

FIG. 2-4-14 Principle diagram of the rig break-proof device.

FIG. 2-4-15 Rig break-proof device is installed on site.

Currently, this equipment is not installed in any other drilling crew except in the original Sichuan Petroleum Bureau in China.

Remedy

The BOP/gear break-proof safety device of the rig brake (drill rig break-proof device) is suitable for all pneumatic brake and eddy brake system drilling rigs with BOPs. This equipment is made of a two-location, three-passageway pneumatic reversing valve, button valve, and shuttle valve. One side of it is connected with a compressed air pipe of the BOP driller's console, and the other is parallel to the crown protector and pneumatic brake pipe. It can brake the drum while the pipe-ram BOP is closed. Even if the driller forgets that the pipe-ram BOP has been closed and tries to lift the drill string, the brake lever cannot be raised. In this way, it will prevent the hydraulic BOP from being damaged, the wellhead from being pulled off, the drill pipe from breaking, or the derrick from falling, causing a severe accident (Figs. 2-4-14, 2-4-15).

Chapter 3

Common Hidden Dangers and Remedies of Choke Lines

Common hidden dangers of the choke line's gate valve:

- The No. 3 gate valve is a hydraulic valve.
- The No. 3 manual valve is usually switched off.
- An obstacle at the hand wheel of a valve obstructs the turning of the hand valve.
- The gate valve has no hand wheel.
- The hand wheel faces toward the drawworks.

Common hidden dangers of choke lines:

- The length of the choke line is over 7 m without being fixed.
- The choke line should only use a hard line.
- Thread connecting the flange is used in the choke line.

1. THE GATE VALVE OF CHOKE LINES

Hidden danger: The No. 3 gate valve is a hydraulic valve

Hazard

During the industrial production, if there is a double gate valve collocated, the internal valve usually is used as a spare valve; the outer valve is the valve used daily. In general, start and stop the channel by switching on and off the outer valve; as a result, the outer valve is more easily damaged or stabbed. Once the outer valve is stabbed, we can switch off the internal valve in order to change the outer valve. If the internal valve is stabbed, we should stop the related equipment or cut off the channel and then replace the internal valve.

There are two gate valves on the right of the spool of the BOP's choke manifold. The manual valve is normally in an open state; the shut-in procedures and choke and kill operations are controlled by the hydraulic valve switch, which realizes that the spool and the choke manifold are connected and closed. The third gate valve is the oil-controlled valve, which is frequently switched and easy to damage. When an overflow or blowout occurs, and the No. 3 hydraulic valve is damaged or stabbed, if this valve cannot be replaced in the bearing pressure situation, it may cause an out-of-control blowout (Figs. 3-1-1 to 3-1-4).

In January 2009, a particular oil field had an empty well blowout shut-in, the result of the No. 3 gate valve being stabbed. Killing the operation could not be implemented, so they were forced to take the risk of having an out-of-control blowout and fire by opening the blind-ram BOP to provide wellhead pressure relief, and rushing to replace the No. 3 valve. If the No. 4 valve was stabbed, and No. 3 was just closed (inside the valve), the No. 4 valve is easily replaced.

FIG. 3-1-1 The No. 3 valve is a hydraulic gate valve (A).

FIG. 3-1-2 The No. 3 valve is a hydraulic gate valve (B).

FIG. 3-1-3 The No. 3 valve is a hydraulic gate valve (C).

FIG. 3-1-4 The No. 3 valve is a hydraulic gate valve (D).

Remedy 1

Gate valve No. 3 should be the manual gate valve, which is normally in an open state; the No. 4 gate valve is the hydraulic valve, which is normally in a closed state. When pressure testing, the BOP drills or shut-in because of overflow, through the No. 4 hydraulic valve, to achieve the on or off of the spool and the choke manifold (Figs. 3-1-5 to 3-1-8).

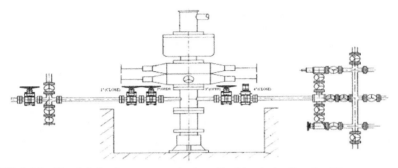

FIG. 3-1-5   Nos. 1, 2, 3 valves are manual valves; No. 4 is a hydraulic valve (A).

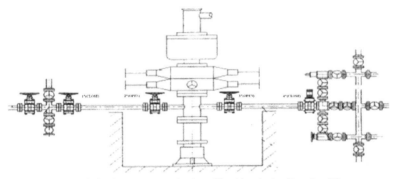

FIG. 3-1-6   Nos. 1, 2, 3 valves are manual valves; No. 4 is a hydraulic valve (B).

FIG. 3-1-7   No. 4 is a hydraulic valve (A).   **FIG. 3-1-8**   No. 4 is a hydraulic valve (B).

Remedy 2

In Figure 3-1-5, the No. 1 and 4 valves are normally closed, and No. 2 and 3 are normally open. By remote control the No. 4 hydraulic valve's switch to achieve choke line and choke manifold pipeline can be connected and closed. If you want to control the relief line and choke manifold connection and cut-off, operators need to enter the substructure of the derrick to open valve No. 1. It is very dangerous for an operator to enter the narrow substructure to operate the No. 1 valve, because of the poor means of escape.

The Chinese industrial standard requests that the first gate valve is the manual gate valve. Some of the overseas oil fields, as shown in Figure 3-1-9, install one hydraulic valve on each side of the spool, which allows the operator to kill the well line connection and cut-off through remote control, eliminating the risk of the operator entering the derrick substructure to operate the No. 1 valve. On some high-risk deep exploratory wells and ultra-deep wells it is recommended to install the device so that the No. 1 and 4 valves are the hydraulic valves. Some oil fields have had this installation, as shown in Figure 3-1-10.

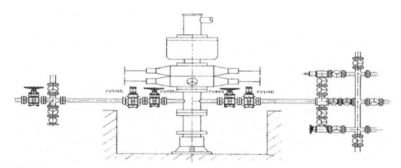

FIG. 3-1-9　Nos. 2, 3 manual valves are normally open; Nos. 1, 4 hydraulic valves are normally closed (A).

FIG. 3-1-10　Nos. 2, 3 manual valves are normally open; Nos. 1, 4 hydraulic valves are normally closed (B).

Hidden danger: The No. 3 manual valve is normally closed

Hazard

When a well needs to be shut off because of an overflow or blowout, the operator must enter the substructure of the derrick, open the third (normally closed) valve, and then close the ram BOP (Fig. 3-1-11). It is inconvenient and dangerous for an operator to enter the narrow substructure to open the No. 3 valve. If the operator cannot open the No. 3 manual valve in time and the condition of the well cannot allow a shut-in by force, it will lead to an out-of-control wellhead.

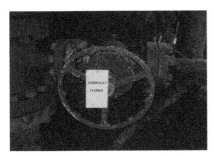

FIG. 3-1-11   No. 3 valve stays in a closed state.   **FIG. 3-1-12**   A blowout and out-of-control well.

In October 2006, an oil field well had a surface casing depth of 129 m, second spudding to a depth of 1294 m, when a blowout occurred. Worried that the No. 3 manual valve was in the off-position, in order to avoid a forced shut-in, making oil and gas escape to the ground, the crew did not dare to close the ram BOP, which led to an out-of-control blowout, as shown in Figure 3-1-12. The manual valve is normally open and the hydraulic valve is in a normally in the off state. If the hydraulic valve was opened and the BOP closed, the well blowout would not have happened.

Remedy

The No. 3 valve is normally open and the No. 4 hydraulic valve is normally closed. The driller can quickly open the remote control valve 4 when needed, to achieve soft shut-in, as shown in Figures 3-1-13 and 3-1-14.

FIG. 3-1-13   No. 3 valve is normally open (A).   **FIG. 3-1-14**   No. 3 valve is normally open (B).

Hidden danger: An obstacle at the valve's hand wheel obstructs the turning of the hand wheel

Hazard

If there is an obstacle at the valve's hand wheel, the hand wheel will not turn or it will be difficult to turn (Figs. 3-1-15, 3-1-16). This situation generally occurs in the spool double-connected valve, with the obstacle on the third valve hand wheel, such as the No. 4 hydraulic valve hydraulic control pipeline or the sewage pipe placed in the cellar or hanging pump's hanging rope (chain). In Figure 3-1-15, the hydraulic control line and the hydraulic valve's interface uses the straight connector; the pipeline position cannot be adjusted, so it is difficult to avoid the third valve hand wheel, which will impede its rotation.

FIG. 3-1-15 Fluid control lines prevent the hand wheel rotation.

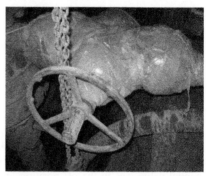

FIG. 3-1-16 Pump rotating prevents the hand wheel rotation.

FIG. 3-1-17 Use a bent sub.

FIG. 3-1-18 Use a rotatable bent sub.

Remedy

Nothing should obstruct the manual valve hand wheel from turning. For the spool double-connected valve, use a 90° to 120° bent sub between the hydraulic-drive valve and the hydraulic pipeline so that the hydraulic pipeline can avoid the third valve hand wheel. This also helps protect the hydraulic pipeline from being bent and damaged by force (Figs. 3-1-17, 3-1-18).

Hidden danger: A gate doesn't have a hand wheel

Hazard

It is not convenient to use a wrench to switch a gate. Also, if the wrench is too short, the on-off torque will be small, or if the wrench is too long, the on-off torque will be too large; both situations can damage the valve. If the gate valve does not have a hand wheel, then the strength will concentrate in the spanner's root, and the spanner will easily break off from the root. In Figure 3-1-19 the gate valve doesn't have a hand wheel, and in Figure 3-1-20, the hand wheel is not installed according to standards.

FIG. 3-1-19 The gate valve lacks a hand wheel.

FIG. 3-1-20 The gate valve hand wheel is not installed according to standard.

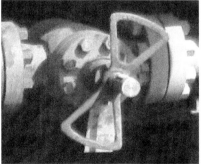

FIG. 3-1-21 Standard hand wheel (A).

FIG. 3-1-22 Standard hand wheel (B).

Remedy

The valve should use the standard hand wheel, with a diameter not less than 450 mm and not over the length of the valve structure (Figs. 3-1-21, 3-1-22).

Hidden danger: The hand wheel faces toward the drawworks

Hazard

The space between the BOP and the drawworks is narrow, the environment is bad, the channel is impeded, and going in or out is not convenient. The spool valve's hand wheel faces the drawworks, which is not good for switching the valve fast, and is not conducive to protect the operator (Figs. 3-1-23, 3-1-24).

FIG. 3-1-23 Valve hand wheel faces the draw-works (A).

FIG. 3-1-24 Valve hand wheel faces the drawworks (B).

Remedy

To facilitate the operation, to provide a good working environment for the operator, and to facilitate evacuation, the valve hand wheel should face the direction of the V-door (Figs. 3-1-25, 3-1-26).

FIG. 3-1-25 Valve hand wheel faces the derrick V-door (A).

FIG. 3-1-26 Valve hand wheel faces the derrick V-door (B).

2. RELIEF PIPE

Hidden danger: The choke line length surpasses 7 m and is not fixed

Hazard

To suppress vibration of the pipeline and to ensure the sealing performance of joints, *SY/T 5964-2006* states that "choke line length should be fixed firmly if over 7 m." In Figure 3-2-1, the pipeline is not fixed but its length is more than 7 m.

FIG. 3-2-1 Relief pipe length is more than 7 m without fixing.

FIG. 3-2-2 Relief pipe length is more than 7 m with the middle part fixed.

Remedy

If the choke lines length is more than 7 m, the middle part should have a fixed point, as shown in Figure 3-2-2.

Defect: The choke lines can only use a hard-line pipeline

Hazard

Using a hard pipeline to connect the spool and choke and kill well manifolds requires a high central deviation. When the choke and kill well manifolds flange and the spool flange bolt holes center deviation does not surpass the aperture's 5%, then the bolt can penetrate freely. In general, the choke and kill well manifold's height can be adjusted to fit the centralization. If the throttle kill manifold pier base's height is adjustable, the connection will be convenient, otherwise, it depends upon the amount of lime soil under the pier base to fit the connection requests. However, this way is inconvenient, unreliable, and easy to jitter (Figs. 3-2-3, 3-2-4).

Using steel pipe as a relief pipe is very inconvenient. It will take several hours for roughnecks to complete the connection.

FIG. 3-2-3 Fill wood, cement brick are under the pier base.

FIG. 3-2-4 Fill sand, soil are under the pier base.

Remedy

Use high-pressure fire resistant hose as a relief line, through a flexible hose to connect the spool and throttle manifold together, so that even if the throttle kill well manifold's axis deviates a great deal from the axis of the spool, connection is very easy.

All offshore drilling uses the hose as the relief pipeline; moreover, some countries' on-land drilling also uses the hose as the relief pipeline (Figs. 3-2-5 to 3-2-8).

FIG. 3-2-5 Offshore drilling uses flexible relief pipeline.

FIG. 3-2-6 Overseas drilling uses flexible relief pipeline (A).

FIG. 3-2-7 Overseas drilling uses flexible relief pipeline (B).

FIG. 3-2-8 Overseas drilling uses flexible relief pipeline (C).

The industrial standard states "the relief pipeline connects to throttle manifold and kill well manifolds with rated operating pressure equal or bigger than 35 MPa should use metallic material." If the rated operating pressure is smaller than 35 MPa, we can use the hose as the choke lines (Figs. 3-2-9, 3-2-10).

FIG. 3-2-9 Domestic drilling uses flexible relief pipeline (A).

FIG. 3-2-10 Domestic drilling uses flexible relief pipeline (B).

Defect: The choke lines uses a threaded flange

Hazard

If you use a threaded joint flange, the bearing pressure capacity is lower than with the welded flange. The threaded joint flange is suitable for the low-pressure, small-diameter pipeline, however the choke lines is for the high-pressure, large-diameter pipeline (Figs. 3-2-11 to 3-2-14).

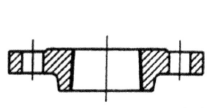

FIG. 3-2-11   Threaded joint flange sectional drawing.

FIG. 3-2-12   Threaded joint flange (A).

FIG. 3-2-13 Threaded joint flange (B).

FIG. 3-2-14 Threaded joint flange (C).

Remedy

There are two methods to connect the flange: one is the threaded joint flange and the other is the welded flange. The relief line belongs to the high-pressure, large-diameter pipeline; as a result, it should use the welded flange instead of the screw joint flange (Figs. 3-2-15 to 3-2-18).

FIG. 3-2-15 Flat welding flange.

FIG. 3-2-16 Friction welding flange with neck.

FIG. 3-2-17 Welded flange (A).

FIG. 3-2-18 Welded flange (B).

Chapter 4

Common Hidden Dangers and Remedies for Choke and Kill Manifolds

Common hidden dangers for throttle and well killing manifolds:

- Obstacles impede the choke manifold valve hand wheel rotation.
- The throttle and well kill manifold five-way pressure gauge control valve does not use a sign to indicate the on-off state.
- Choke and kill manifolds without a pier base or foundation are not stable.
- Choke and kill manifolds do not use an adjustable height pier base.
- The gate valve on-off sign is not legible.

Common hidden dangers of pressure gauges:

- Choke and kill manifolds do not have a small-range pressure gauge.
- The high pressure gauge range for the choke and kill manifolds does not match the rated pressure for manifolds.
- The pressure gauge has a tilted or inverted installation.
- The back of the gauge dial is to the manual throttle valve operator.

1. THROTTLE AND WELL KILLING MANIFOLDS

Hidden danger: Obstacles hinder the choke manifold gate valve hand wheel rotation

Hazard

Without a five-way pressure gauge control valve in the throttle manifold, if the manifold pressure gauge cut-off valve is in the wrong position, it will obstruct the cut-off valve hand wheel and prevent it from revolving (Figs. 4-1-1, 4-1-2).

Remedy

Adjust the pressure gauge cut-off valve's position, and avoid obstructing the gate valve hand wheel with the cut-off valve. In addition, the operator can guide a pressure tube to elevate the pressure gauge and cut-off valve to avoid the spool gate valve hand wheel (Figs. 4-1-3, 4-1-4).

FIG. 4-1-1 The cut-off valve hand wheel obstructs the gate valve hand wheel (A).

FIG. 4-1-2 The cut-off valve hand wheel obstructs the gate valve hand wheel (B).

FIG. 4-1-3 The cut-off valve hand wheel does not hinder the gate valve hand wheel (A).

FIG. 4-1-4 The cut-off valve hand wheel does not hinder the gate valve hand wheel (B).

Defect: The choke and kill manifolds five-way pressure gauge control valve does not indicate the on-off state with a hanging plate

Hazard

SY/T 5964-2006 requires "choke manifold and kill manifold valve should follow the way of Fig. 7 and Fig. 11 to list number and indicate on and off states." Unfortunately, this standard does not name or number the throttle and well killing manifold's five-way pressure gauge control valve, nor does it list the on-off states when awaiting orders. Sometimes, operators fail to list the on-off states of the pressure gauge control valve with a hanging plate and this gate valve's on-off state is inconsistent when awaiting orders (Figs. 4-1-5, 4-1-6).

FIG. 4-1-5 The throttle manifold pressure gauge control valve has no hung plate.

FIG. 4-1-6 The well killing manifold pressure gauge valve has no hung plate.

Remedy

The on-off states of the five-way-pressure gauge control valve should be listed, and this valve should be switched on during the await orders working condition (Figs. 4-1-7, 4-1-8). The current industrial standard does not number this valve, which is convenient for maintenance and management. This valve is not considered in the same plane with other valves, and as a result, the choke manifold pressure gauge control valve is numbered J0 and the well killing manifold pressure gauge control valve is numbered Y0.

FIG. 4-1-7 The plate is hung at the pressure gauge control valve (A).

FIG. 4-1-8 The plate is hung at the pressure gauge control valve (B).

Hidden danger: The choke and kill manifolds do not have a pier base or the foundation is not firm

Hazard

If the choke and kill manifolds do not have a pier base or the foundations are not firm, the manifold stability will be poor. In the process of throttle or well killing, the intensity of the manifold's shaking affects the manifold flange joint's seal and can cause the seal to lose efficacy (Figs. 4-1-9, 4-1-10).

FIG. 4-1-9 The choke manifold does not have a substructure; it is hung in the air.

FIG. 4-1-10 The kill-line manifold substructure footing is not firm.

Remedy

In order to improve the throttle and well killing manifold's stability, the substructure should be placed on a solid base (Fig. 4-1-11). The preferred method is to place the throttle and well killing manifolds on a cement bar (Fig. 4-1-12).

FIG. 4-1-11 The well killing manifold is placed on the tamped ground.

FIG. 4-1-12 The throttle manifold is placed on the concrete strip.

Defect: The throttle and well killing manifolds do not use an adjustable height pier base

Hazard

At present, almost all Chinese on-land drilling uses steel pipeline as the relief line. The choke line and the choke manifold and throttle are connected through a flange, and the axis deviation of the flange joint should be very small, otherwise it is difficult to buckle a steel ring into the flange groove. Using a nonadjustable height throttle manifold pier base, in order to make sure the relief line is connected to the throttle and well killing manifolds, we can only fill rock and soil under the pier base (Figs. 4-1-13, 4-1-14) to adjust the throttle manifold height. This is inconvenient and time-consuming; moreover, the base foundation is not firm. In Figures 4-1-15 and 4-1-16, the manifold pier base height is not adjustable and the installation is very difficult.

FIG. 4-1-13   Fill stones or wood bricks under the throttle manifold pier base.

FIG. 4-1-14   Fill loose sand under the throttle manifold pier base.

FIG. 4-1-15   Nonadjustable pier base height (A).

FIG. 4-1-16   Nonadjustable pier base height (B).

Remedy 1

When the throttle manifold and relief line are connected, the throttle manifold pier base height should be adjustable to improve the stability of the manifold (Figs. 4-1-17, 4-1-18).

FIG. 4-1-17   Adjustable throttle manifold pier   FIG. 4-1-18   Adjustable throttle manifold pier
base height (A).   base height (B).

Remedy 2

If the relief line is a high-pressure fire resistant hose, whether the throttle manifold substructure height is adjustable or not, the manifold installation is simple,
convenient, and fast.

Defect: The valve's on-off state is not legible on the nameplate

Hazard

The purpose of hanging a sign-plate at the gate valve is to provide clear,
convenient, rapid, and accurate identification of the valve's on-off status.
If the writing on the sign-plate is too small, poorly visible at night, or contaminated with fluid (this is often the case), the sign-plate will not be fully clear
(Figs. 4-1-19, 4-1-20).

FIG. 4-1-19   The same color sign-plate is not   FIG. 4-1-20   The same color sign-plate is not
clear (A).   clear (B).

The throttle manifold valve hand wheel is the same color, and the daily state
of the valve is not identified by different colors (Figs. 4-1-21, 4-1-22).

Remedy 1

For convenience, and for fast, clear, and accurate identification of the valve's on-off state, the sign-plates should use particular colors: red should indicate that the valve is closed, and green should indicate that the valve is open (Figs. 4-1-23, 4-1-24).

FIG. 4-1-21   Valve hand wheel of the same color (A).

FIG. 4-1-22   Valve hand wheel of the same color (B).

FIG. 4-1-23   Clear two-color identification plate (A).

FIG. 4-1-24   Clear two-color identification plate (B).

Remedy 2

An alternative method of marking the gate valve is to paint the gate switch using a state nameplate (Figs. 4-1-25, 4-1-26). The valve hand wheel should be painted in two colors, red and green (or blue). Normally red means the valve is closed, and green (or blue) means it is open. If the valve state changes, it can be listed elsewhere.

FIG. 4-1-25 Hand wheel color-coded the different on-off states (A).

FIG. 4-1-26 Hand wheel color-coded the different on-off states (B).

Remedy 3

A protective cap with rings on the valve stem can be used to identify the gate's on-off state: green is on and red is off. This method is simple, convenient, intuitive, and easy to operate. If the valve status changes, just remove the old rings and put on different color rings. This method is strongly recommended (Figs. 4-1-27, 4-1-28).

Colored reflective material should also be used for night observations.

FIG. 4-1-27 Using color-coded rings to distinguish the valve's on-off state (A).

FIG. 4-1-28 Using color-coded rings to distinguish the valve's on-off state (B).

2. PRESSURE GAUGE

Hidden danger: Throttle and well killing manifolds are without a low-range pressure gauge

Hazard

Pressure measured between the pressure gauge's upper limit of $\frac{1}{3}$ to $\frac{2}{3}$ (or $\frac{1}{4}$ to $\frac{3}{4}$) is more accurate. When the pressure is less than the upper limit, the absolute error and visual observation results in a huge deviation on the high-pressure gauge, which will impact pressure reading accuracy (Figs. 4-2-1 to 4-2-4).

Remedy

FIG. 4-2-1   Pressure measured between the measuring range upper limit of ⅓ to ⅔ (A).

FIG. 4-2-2   Pressure measured between the measuring range upper limit of ⅓ to ⅔ (B).

FIG. 4-2-3   The throttle manifold does not have a low-range pressure gauge.

FIG. 4-2-4   The well killing manifold does not have a low-range pressure gauge.

The press classes commonly used for throttle and well killing manifolds are 21 MPa, 35 MPa, 70 MPa, and 105 MPa; the corresponding pressure gauges are 30 MPa, 50 MPa, 100 MPa, and 150 MPa. When pressure is below the full scale of one-third (10 MPa, 16 MPa, 33 MPa, and 50 MPa), the reading has huge errors. Selecting the pressure gauge fully as the pressure gauge's range, which may need several gauges, especially for 70 MPa and 105 MPa manifolds. Aim for operational and practical, as throttle and well killing manifolds may be installed at both high- and low-range pressure gauges (Figs. 4-2-5, 4-2-6).

The range of the low-pressure gauge installed on throttle and well killing manifolds is greater than one-third of the high-pressure gauge's range. For a 21 MPa manifold, its high-pressure gauge is 30 MPa and low-pressure gauge is 10 MPa; for a 35 MPa manifold, its high-pressure gauge is 50 MPa and low-pressure gauge is 16 MPa; for a 70 MPa manifold, its high-pressure gauge is 100 MPa and low-pressure gauge is 25 MPa; for a 105 MPa manifold, its high-pressure gauge is 150 MPa and low-pressure gauge is 35 MPa.

FIG. 4-2-5 High- and low-range pressure gauge of throttle manifold.

FIG. 4-2-6 High- and low-range pressure gauge of well killing manifolds.

FIG. 4-2-7 Install a cut-off valve before the throttle manifold low-pressure gauge, which is normally closed.

FIG. 4-2-8 Install a cut-off valve before the well and control manifold low-pressure gauge, which is normally closed.

A five-way pressure gauge control valve is equipped with two pressure gauges. A cut-off valve should be installed before the low-pressure gauge; the cut-off valve is normally closed (Figs. 4-2-7, 4-2-8). It is not necessary to install a cut-off valve before a high-pressure gauge. Use a five-way pressure gauge control valve to control the closing and connection.

Defect: The throttle and well killing manifold high-pressure gauge's range does not match the rated pressure

Hazard

It is a common phenomenon that a pressure gauge's range does not match the rated pressure; for example, a 21 MPa manifold installed a pressure gauge with 50 MPa or 60 MPa and a 35 MPa manifold installed a 40 MPa or 60 MPa pressure gauge. If the gauge's range does not fit the rated pressure, the absolute error and bias of visual observation will be huge, which impacts pressure reading accuracy (Figs. 4-2-9, 4-2-10).

FIG. 4-2-9   High-pressure gauge of 35 MPa throttle manifold (A).

FIG. 4-2-10   High-pressure gauge of 35 MPa throttle manifold (B).

Remedy

In order to obtain more accurate pressure data from a well, the rated pressure of a manifold should be two-thirds or three-quarters of the pressure gauge's full range. The recommend selection of gauge is that:

- 21 MPa throttle and well killing manifolds need a range of 30 MPa high-pressure gauge
- 35 MPa throttle and well killing manifolds need a range of 50 MPa high-pressure gauge (Fig. 4-2-11)

- 70 MPa throttle and well killing manifolds need a range of 100 MPa high-pressure gauge (Fig. 4-2-12)
- 105 MPa throttle and well killing manifolds need a range of 150 MPa high-pressure gauge

FIG. 4-2-11 A high-pressure gauge's range is 50 MPa.

FIG. 4-2-12 A high-pressure gauge's range is 100 MPa.

Hidden danger: The pressure gauge is installed either tilted or inverted

Hazard

A pressure gauge that is installed either tilted or inverted is not in the normal position, which is not convenient to reading, can indicate a large error, shortens the lifetime of the gauge, and reduces anti-seismic performance (Figs. 4-2-13, 4-2-14).

FIG. 4-2-13 Pressure gauge installed inverted (A).

FIG. 4-2-14 Pressure gauge installed inverted (B).

Remedy

The pressure gauge must be installed vertically upward, angled generally at no more than 30°; installing it inverted is not allowed (Figs. 4-2-15, 4-2-16).

FIG. 4-2-15 Vertically installed pressure gauge (A).

FIG. 4-2-16 Vertically installed pressure gauge (B).

Defect: The back of the pressure gauge dial is to the operator of the manual choke valve

Hazard

If the back of the pressure gauge dial installed on the throttle manifold is to the manual throttle valve operator, it is difficult to read the pressure (Figs. 4-2-17 to 4-2-20).

FIG. 4-2-17 The pressure gauge's direction is bad for reading (A).

FIG. 4-2-18 The pressure gauge's direction is bad for reading (B).

FIG. 4-2-19 The pressure gauge in back of the manual throttle valve.

FIG. 4-2-20 The pressure gauge faces the hydraulic throttle valve.

Remedy

When installing the pressure gauge, the dial should face the operator, and face the two gauges' dials in the same direction. If the direction of the two dials is slightly different but it is convenient to read the pressure, it is also acceptable (Figs. 4-2-21, 4-2-22).

FIG. 4-2-21 Direction of the pressure gauge's dial is inconformity (A).

FIG. 4-2-22 Direction of the pressure gauge's dial is inconformity (B).

The No. 1 and 4 valves are throttle valves, so the pressure gauge should face the direction of the derrick substructure; that way the operator can easily read the pressure for either throttle valve (Figs. 4-2-23, 4-2-24). If the No. 1 valve is a hydraulic throttle valve and the No. 4 valve is a manual throttle valve, the pressure gauge should face the derrick substructure or the No. 4 manual valve, but should not face the hydraulic throttle valve.

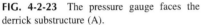

FIG. 4-2-23 The pressure gauge faces the derrick substructure (A).   **FIG. 4-2-24** The pressure gauge faces the derrick substructure (B).

In order to make the gauge dial face in a certain direction, several adjustments to the anneal red copper pad, lead pad height, or sealing tape around threads might be necessary. The manager should allow the gauge dial to side-face the operator and allow the direction of two gauges to be slightly different when these do not affect reading the pressure.

The connection's seal is more important than the direction of the gauge dial. Moreover, facing the gauge dial away from the operator should not be considered a hidden danger.

Chapter 5

The Common Hidden Dangers and Remedies for Drilling Fluid Recovery Pipeline

Common hidden dangers of the drilling fluid recovery pipeline:

- The drilling fluid recovery pipeline's bent sub angle is less than 120°.
- The drilling fluid recovery pipeline's bent sub is not made of cast steel.
- The flanged joint is sealed by union or asbestos pad.
- The flange face is not parallel.
- The bent sub lies in the air.
- The drilling fluid recovery pipeline is made of hard pipeline.
- The pipeline goes into another tank behind the degasser.

Common hidden dangers of fixing the drilling fluid recovery pipeline:

- The drilling fluid recovery pipeline's bent sub is not fixed.
- The middle section of the over-length drilling fluid recovery pipeline is not fixed.
- The fixing of the exit port is not secure.
- The outlet fixing point of the drilling fluid recovery pipeline is filled by a piece of wood.

1. DRILLING FLUID RECOVERY LINE

Hidden danger: The drilling fluid recovery pipeline's bent sub angle is less than 120°

Hazard

The drilling fluid recovery hard pipeline connected to the No. 8 valve needs at least two bent subs; generally, for the drilling fluid to reach the circulate tank, we use three bent subs. The angle of the bent sub is smaller; the erosion of the drilling fluid to the bent sub is more difficult. When the angle of the bent sub is less than 120°, the ability of the bent sub's anti-erosion is obviously weakened (Figs. 5-1-1 to 5-1-4).

FIG. 5-1-1   The angle of the bent sub is about 90° (A).

FIG. 5-1-2   The angle of the bent sub is about 90° (B).

FIG. 5-1-3   The angle of the bent sub is about 80°.

FIG. 5-1-4   The angle of the bent sub is about 100°.

Remedy

The angle of the bent sub used in the drilling fluid recovery pipeline should not be less than 120°; moreover, we can also use the erosion resistant right-angle irrigation lead bent sub (Figs. 5-1-5 to 5-1-8).

FIG. 5-1-5   The angle of the bent sub is larger than 120°.

FIG. 5-1-6   The right-angle irrigation leads to the bent sub.

FIG. 5-1-7   The angle of the bent sub at the exit is larger than 120° (A).

FIG. 5-1-8   The angle of the bent sub at the exit is larger than 120° (B).

Hidden danger: The drilling fluid bent sub is not made of cast steel

Hazard

A bent sub made by either welded or curved steel pipe will have thin wall thickness and poor erosion resistance. Even if a cast steel bent sub is used, under the erosion of high density drilling fluid, the wash out phenomenon appears (Figs. 5-1-9 to 5-1-14).

FIG. 5-1-9   Butt welding noncast steel bent sub.

FIG. 5-1-10   Curved steel pipe.

FIG. 5-1-11   Butt welding cast steel bent sub (A).

FIG. 5-1-12   Butt welding cast steel bent sub (B).

FIG. 5-1-13 The recovery pipeline bent sub stabbed in a blowout.

FIG. 5-1-14 The stabbed non-cast steel bent sub.

Remedy

The material used for drilling fluid recovery pipeline in a bent sub should be cast steel and the angle should not be less than 120°. Moreover, we can also use the erosion resistant right-angle irrigation lead bent sub (Figs. 5-1-15 to 5-1-18).

FIG. 5-1-15 The angle of the cast steel bent sub is larger than 120° (A).

FIG. 5-1-16 The angle of the cast steel bent sub is larger than 120° (B).

FIG. 5-1-17 The angle of the cast steel bent sub is larger than 120° (C).

FIG. 5-1-18 The angle of the cast steel bent sub is larger than 120° (D).

Hidden danger: The drilling fluid recovery pipeline is connected by a union or flange with an asbestos sealing pad

Hazard

In the process of choke and kill operation, the pressure in the drilling fluid recovery pipeline drops rapidly and the pipeline shakes acutely. The union joint is convenient but it is a movable joint; if it gets shaken, it can get free and pierce, which will lead to the seal losing efficacy (Figs. 5-1-19, 5-1-20).

FIG. 5-1-19   Union joint (A).

FIG. 5-1-20   Union joint (B).

The pressure capacity of the flange sealed by asbestos is not over 5 MPa. Although the exit of the drilling fluid recovery pipeline connects to the air, the pressure of the pipeline's entrance approaches the pressure at the rear of the throttle valve, where the pressure is over 5 MPa. As a result, the flange sealed by asbestos cannot satisfy the pressure capacity of the pipeline. In the process of choke and kill operation, the flange will be stabbed (Figs. 5-1-21, 5-1-22).

FIG. 5-1-21   Flanged joint sealed by asbestos.

FIG. 5-1-22   The pipeline uses asbestos to seal the flanged joint.

Remedy

To keep the performance of the drilling fluid recovery pipeline joint's pressure capacity, we should use a high-pressure capacity tenon groove seal flanged joint; union or asbestos is not allowed to seal flanged joints (Figs. 5-1-23, 5-1-24).

FIG. 5-1-23   High-pressure flanged joint.   **FIG. 5-1-24**   Tenon groove flanged joint.

Hidden danger: The flanged surface of the drilling fluid recovery pipeline is not parallel

Hazard

If the flanged surface is not parallel at the flanged joint, it is hard to ensure the joint's seal (Fig. 5-1-25).

FIG. 5-1-25   Flanged surface is not parallel at   **FIG. 5-1-26**   Flanged surface is parallel at the
the flanged joint.   flanged joint.

Remedy

Generally, the joint does not take the test of pressure capacity at the location of the drilling fluid recovery pipeline; as a result, it cannot test the sealing performance of the pipeline. In order to keep the sealing performance of the joint, we should only keep the flanged surface parallel at the joint, and the deviation should not be larger than the flanged joint's outer diameter of 1.5%, which should not be larger than 2 mm (Fig. 5-1-26). Tightening the bolt by force to remove the oblique of the flanged joint is not allowed. Moreover, connecting by force to remove the deviation between flanged joints is not allowed.

Hidden danger: The bent sub of the drilling fluid recovery pipeline lies in the air

Hazard

The bent sub of the drilling fluid recovery pipeline that lies in the air will be hard to fix, and the stability of the pipeline is poor (Figs. 5-1-27, 5-1-28).

FIG. 5-1-27   The bent sub lies in the air (A).   **FIG. 5-1-28**   The bent sub lies in the air (B).

Remedy

The bent sub should be located where it is easy to fix (Figs. 5-1-29, 5-1-30). We can install a bent sub near the No. 8 valve and lead the pipeline to the tank surface. After that, install a bent sub on the top of the tank, so that this bent sub can be installed on the surface and can lead the pipeline to the No. 1 tank horizontally.

FIG. 5-1-29   The bent sub lies on the ground.   **FIG. 5-1-30**   The bent sub is easy to fix.

Defect: Hard pipeline is used as the drilling fluid recovery pipeline

Hazard

When the throttle manifold is in place, the No. 8 valve's place is fixed and the place of connection between the outlet of the recovery pipeline and the inlet of the tank is fixed. It is difficult to use hard pipeline to connect these two fixed points. When they cannot be connected, the only method is to change the position of the recovery pipeline outlet (Figs. 5-1-31, 5-1-32).

FIG. 5-1-31   The screws of adjusting the outlet place lie on the rim of the tank.

FIG. 5-1-32   The U-shaped clamp holes lie on the wall of the tank.

Each well's trip tank and borehole distance and location are not the same. If the hard pipeline is used, the pipeline's temporary outlet should be changed for every well. That not only is more work, but it is difficult to connect and cannot ensure the temporary welding joint's stability (Figs. 5-1-33, 5-1-34).

FIG. 5-1-33   The welding joint is not fastened.

FIG. 5-1-34   The welding joint drops.

Remedy

Use the high-pressure fire resistant soft pipeline as the recovery pipeline. It is convenient, saves effort, and fastens the pipe on the tank. We can fix the outlet bent sub on the tank surface, and then connect the soft pipeline's ends to the No. 8 valve and outlet bent sub. Although the distance and place between trip tank and borehole are different from the other wells, the pipeline connection is very convenient. Moreover, using soft pipeline can reduce the number of bent subs; only one or two bent subs are needed to raise the pipe to the tank (Figs. 5-1-35 to 5-1-38).

FIG. 5-1-35   Use high-pressured soft pipeline as the recovery pipeline (A).

FIG. 5-1-36   Use high-pressured soft pipeline as the recovery pipeline (B).

FIG. 5-1-37   The outlet of the pipeline is fixed on the tank (A).

FIG. 5-1-38   The outlet of the pipeline is fixed on the tank (B).

Hidden danger: The drilling fluid recovery pipeline goes into another tank behind the degasser

Hazard

In the process of choke and kill manifold, the drilling fluid recovery pipeline goes into another tank behind the degasser. The drilling fluid returned from the recovery pipeline contains gas, and if this kind of drilling fluid goes into the circulating ditch without passing the degasser first, it will increase the pressure, losing balance in the well after it is pumped in (Figs. 5-1-39, 5-1-40). Moreover, gas-bearing drilling fluid that doesn't pass through the degasser is heavy and will lead to natural gas gathering in the tank area.

FIG. 5-1-39 Recovery pipeline goes into another tank behind the degasser (A).

FIG. 5-1-40 Recovery pipeline goes into another tank behind the degasser (B).

Remedy

The outlet of the drilling fluid recovery pipeline lies between the degasser and shale shaker. The gas-bearing drilling fluid should pass through the degasser first, and then into the circulating ditch (Figs. 5-1-41, 5-1-42).

FIG. 5-1-41 The outlet of the recovery pipeline lies in front of the degasser (A).

FIG. 5-1-42 The outlet of the recovery pipeline lies in front of the degasser (B).

2.   FIXING THE DRILLING FLUID RECOVERY PIPELINE

Hidden danger: The corner of the drilling fluid recovery pipeline is not fixed

Hazard

If the bent sub of the recovery pipeline is not fixed, it will not control the pipeline from shaking, which will cause the flanged joint's seal to lose efficacy (Figs. 5-2-1, 5-2-2).

FIG. 5-2-1   Bent sub is not fixed (A).

FIG. 5-2-2   Bent sub is not fixed (B).

Remedy

The bent sub of the drilling fluid recovery pipeline should be fixed by a footing. This kind of bent sub always stays in a nonhorizontal position, and as a result, fixing the front and back of the bent sub is difficult. If the footing is fixed at the back of the bent sub it will lack support and the stability will be poor. Under the circumstances, it is not necessary to fix the back, but the front part should be fastened. Moreover, we can also use double-footing to fix it (Figs. 5-2-3, 5-2-4).

FIG. 5-2-3   The front of the bent sub is fixed by a footing (A).

FIG. 5-2-4   The front of the bent sub is fixed by a footing (B).

Hidden danger: The middle section of the over-length drilling fluid recovery pipeline is not fixed

Hazard

If the middle section of the over-length drilling fluid recovery pipeline is not fixed, it is difficult to control the pipeline from shaking (Fig. 5-2-5).

FIG. 5-2-5 The middle section of the over-length drilling fluid recovery pipeline is not fixed.

FIG. 5-2-6 The middle section of the over-length drilling fluid recovery pipeline is fixed by a footing.

Remedy

There are no industry-standard rules for pipeline fixing methods. As a result, reference the fixing method for the relief line: fix the middle of the drilling fluid recovery pipeline if its length is over 7 m (Figs. 5-2-6 to 5-2-8).

FIG. 5-2-7 The middle part of the recovery pipeline is fixed by a land bolt.

FIG. 5-2-8 The middle part of the recovery pipeline is fixed by a footing.

Hidden danger: The outlet of the drilling fluid recovery pipeline is not fastened

Hazard

The pressure of fluid from the well will reduce quickly when it reaches the outlet of the recovery pipeline, which will cause the pipeline to shake acutely, with the outlet taking the maximum force. If the outlet is not fastened, it will crash into the tank and produce sparks, which could lead to an explosion or fire.

Figures 5-2-9 to 5-2-14 show examples of unfastened recovery pipeline outlets. In Figure 5-2-9, the bolts of the gland are welded on the mobile gangway; the gangway is connected to the tank by a hinge that is not rigidly fastened, so it is not firm. Figures 5-2-10 to 5-2-12 show that the welding area is small and unfastened. In Figure 5-2-14, steel wire cannot control the shaking of the outlet.

FIG. 5-2-9 Outlet is fixed on the gangway.

FIG. 5-2-10 Fixing of a welding screw (A).

FIG. 5-2-11 Fixing of a welding screw (B).

FIG. 5-2-12 Fixing of a welding thin steel pipe.

FIG. 5-2-13   Welding joint is broken.   **FIG. 5-2-14**   Fix the outlet with steel wire.

Remedy

The recovery pipeline should connect the tank in a place that will be good to fix the outlet. The fixing point of the outlet should have reinforcing ribs on the tank surface and should increase the welding area of the fixing piece and tank surface while welding (Figs. 5-2-15, 5-2-16). Using a U-shaped bolt across the rim of the drilling fluid tank to fix it is one method (Figs. 5-2-17, 5-2-18).

FIG. 5-2-15   Fixing piece is welded on the rim of the recovery tank.

FIG. 5-2-16   Fixing piece is welded on the reinforcing ribs of the tank.

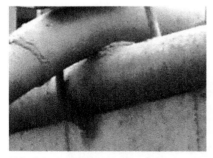

FIG. 5-2-17   Using a U-shape bolt across the rim to fix (A).

FIG. 5-2-18   Using a U-shape bolt across the rim to fix (B).

Hidden danger: The outlet fixing point of the drilling fluid recovery pipeline is filled by a piece of wood

Hazard

If a piece of wood fills the outlet gland of the recovery pipeline, it will be crushed after undergoing force, which will lead to gland slack, and the fixing will lose efficacy (Figs. 5-2-19, 5-2-20).

FIG. 5-2-19 Filling the outlet under the gland with a piece of wood.

FIG. 5-2-20 Filling the outlet under the pipeline with a piece of wood.

Remedy

To ensure that the outlet of the recovery pipeline is fastened, filling the outlet with wood brick under the gland or pipeline is not allowed. Instead, the gland should press against the pipeline tightly (Figs. 5-2-21, 5-2-22).

FIG. 5-2-21 Fixed tightly (A).

FIG. 5-2-22 Fixed tightly (B).

Chapter 6

Common Hidden Dangers and Remedies of Relief Pipe Installation

Common hidden dangers of relief pipe installation include:

- The length of the relief pipe is shorter than the stipulated length of the oil field.
- The relief pipe outlet is higher than the inlet.
- Other devices are next to the relief pipe outlet.
- The relief pipe doesn't have a pipe bridge or the pipe bridge is unable to protect the relief pipe from being run over.
- The drill pipe is used as a relief pipe.
- The sub/interface of the relief pipe did not fasten well.
- The relief pipe passes through the mud waste/reserve pit.
- The target side of the 90° buffer bent sub filled with lead faces the outlet of the relief pipe.

Common hidden dangers of fixed relief pipe include:

- The relief pipe fixed distance is over 15 m.
- The turnings of the relief pipe are not fixed or are fixed too far from the bent sub.
- The fixing of the relief pipe is not firm enough.
- Material from a filling-type pier base leaks.
- The fixing point of the relief pipe outlet is too far away from the outlet, and the pipe outlet is fixed with only one pier base.
- The fixing pier bases are settled on the ground.
- A welded type gland is used.

1. RELIEF PIPE

Hidden danger: The length of the relief pipe is shorter than the stipulated length of the oil field

Hazard

There are many different lengths, according to different types of oil wells in an oil field. While the drilling crew is installing the relief pipe, if the length of the relief pipe is shorter than the stipulated length of the oil field (Figs. 6-1-1, 6-1-2), and the outlet is too close to the equipment, it will fail to ignite at the outlet, and in turn, harmful gas can't be completely burned out so there is potential damage to public security and the environment.

Remedy

The length of the relief pipe should not be shorter than the stipulated length of the oil field. The selection of the length of the relief pipe of each oil field should consider the liquid feature in the borehole, the formation pressure and circumstances, and environment preservation under the premise of security. Those oil wells with formation pressure below 21 MPa and heavy oil wells with depths below 1000 m could have shorter relief pipes; they don't need to be longer than 75 m.

However the length of the relief pipe should exceed 75 m in the high pressure wells, high-sulfur wells, and shallow gas wells.

FIG. 6-1-1 Relief pipe is connected with one single of drill pipe (A).

FIG. 6-1-2 Relief pipe is connected with one single of drill pipe (B).

FIG. 6-1-3 Length of relief pipe meets to release and ignite (A).

FIG. 6-1-4 Length of relief pipe meets to release and ignite (B).

Defect: The relief pipe outlet is higher than the inlet

Hazard

If the relief pipe outlet is higher than the inlet it can cause these problems: first, after testing or releasing, the residual liquid in the pipe is hard to discharge and it can easily consolidate and freeze in the bottom of the pipe and block it, especially in the winter; second, the gushing liquid will flow back to the well site when the amount of the gushing liquid exceeds the volume of the relief pit (Figs. 6-1-5, 6-1-6).

FIG. 6-1-5   Outlet of the relief pipe is higher than the inlet (A).

FIG. 6-1-6   Outlet of the relief pipe is higher than the inlet (B).

Remedy

Make the relief pipe outlet below the inlet under the premise of topography. The relief pipe should be installed at a decline, which will release liquid easily (Figs. 6-1-7, 6-1-8). The expulsion of liquid from a relief pipe is an important factor in the selection of the direction of the well site door.

FIG. 6-1-7   Relief pipe is at a decline (A).

FIG. 6-1-8   Relief pipe is at a decline (B).

Hidden danger: Other devices are next to the relief pipe outlet

Hazard

Flammable and explosive objects are located next to the relief pipe outlet. This creates the risk of igniting harmful and flammable gas when it is released. Any remaining gas is dangerous to the public and environment (Figs. 6-1-9, 6-1-10).

FIG. 6-1-9 Outlet of pipe is near the oil tank, remote console.   **FIG. 6-1-10** Relief pipe is far away, from about 3 m.

Remedy

The industry standards require that the relief pipe outlet should not be less than 50 m from the various devices, and it should be especially far away from flammable and explosive objects (Fig. 6-1-11). The safe distance between the relief pipe outlet and other devices can be judged according to the liquid feature and pressure of the oil wells. Oil wells with a formation pressure below 21 MPa and heavy oil wells with a depth below 1000 m could have a shorter distance between other devices where appropriate; industry rules and regulations determine concrete distances.

The distance between the devices and the flank of the relief pipe outlet can be shortened as well. Usually a distance not less than 30 m between the relief pipe outlet and various devices is considered safe.

A distance less than 50 m between the relief pipe outlet and various devices is acceptable if combustion cells and firewall blocks are around the output port of the relief pipe. The combustion cells and firewall blocks can protect the environment from fire if there are many devices or plants around the relief pipe outlet (Fig. 6-1-12).

FIG. 6-1-11   Outlet of the relief pipe is open.   FIG. 6-1-12   Firewall is at the outlet of the relief pipe.

Hidden danger: The relief pipe doesn't have a pipe bridge or the pipe bridge is unable to protect the relief pipe from being run over

Hazard

There is no pipe bridge in the relief pipe. When vehicles go through the pipe bridge, the relief pipe is compressed by force, affecting the fixing and sealing of pipe, as shown in Figures 6-1-13 to 6-1-16.

FIG. 6-1-13   No pipe bridge where vehicles pass (A).   FIG. 6-1-14   No pipe bridge where vehicles pass (B).

FIG. 6-1-15   Pipe bridge can't prevent the pipe from being run over (A).   FIG. 6-1-16   Pipe bridge can't prevent the pipe from being run over (B).

Remedy

The location for vehicles going across the relief pipe should be set with a pipe bridge, and the bridge should be able to protect the pipe from bending because of being driven over (Figs. 6-1-17, 6-1-18).

FIG. 6-1-17 Pipe bridge protects the relief pipe from being run over (A).

FIG. 6-1-18 Pipe bridge protects the relief pipe from being run over (B).

Defect: The drill pipe is used as the relief pipe

Hazard

Many drilling crews use cheaper, abandoned, old drill pipes as relief pipes (Fig. 6-1-19). Drill pipes rely on sub shoulders to seal. The torque of a hydraulic tong is generally 50 to 100 kN·m. For 5-in old drill pipe joints (sub worn as two grades), the minimum back-up torque is 19.1 kN·m. The back-up torque of a pipe tong or chain tong is generally not over 8 kN·m, which is far less than the torque of the drill pipe. The insufficient torque will lead to seal failure. After the pipe is installed on-site, often 10 MPa pressure can't be maintained.

To increase the back-up torque, a pipe tong or chain tong coordinating reinforcing rod is commonly used in drilling crews. The reinforcing rod can damage the pipe tong or chain tong, however. When connecting the relief pipe, it is very common that pipe tongs or chain tongs get damaged (Fig. 6-1-20).

FIG. 6-1-19 Drill pipe used as a relief pipe.

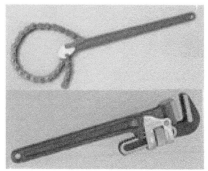

FIG. 6-1-20 Pipe tong and chain tong used in screw-on.

Remedy

Nowadays there are two types of relief pipes: one is a special relief pipe with a tenon-groove sealing flange (steel ring); the other is made of waste drill pipe. The special relief pipe with a tenon-groove sealing flange (steel ring) is able to sustain high pressure and is convenient to connect and fasten. It is recommended, especially in high-kill wells, high sulfur wells, gas wells, and high gas-oil ratio wells (Figs. 6-1-21, 6-1-22).

If a waste drill pipe is used as relief pipe, it should be used only when the formation pressure of the oil well is below 21 MPa.

FIG. 6-1-21   Tenon-groove seal flange (steel ring) relief pipe.

FIG. 6-1-22   Tenon-groove seal (steel ring) flange.

Hidden danger: The interface/sub thread of the relief pipe was not well fastened

Hazard

Drill pipe is sealed by shoulders and the sub is not fastened firmly or tightly enough, causing the seal to fail, being unable to satisfy the demand of 10 MPa (Figs. 6-1-23, 6-1-24).

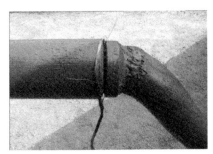

FIG. 6-1-23   Tread is not screwed on tightly; there are surplus threads (A).

FIG. 6-1-24   Tread is not screwed on tightly; there are surplus threads (B).

Remedy

If using the drill pipe as the relief pipe, the thread must be fastened firmly and the back-up torque should be as close to minimum back-up torque as possible.

Hidden danger: Relief pipe is installed through the drilling mud waste pit

Hazard

If the relief pipe is installed through the drilling mud waste pit, it is inconvenient to connect and fix. Installation and dismantling will be difficult. Setting the fixing point will be extremely tough in the middle of the waste pit. If the pipe bridge is not installed, however, the distance of fixing points will exceed 15 m if the fixing point was not set in the middle of the waste pit (Figs. 6-1-25, 6-1-26).

FIG. 6-1-25 The fixing interval is over 15 m.   **FIG. 6-1-26** Relief pipe is installed through a mud waste pit.

In Figure 6-1-27, the relief pipe goes through the waste pit. Although it has a pipe bridge and the distance between the pier bases satisfies the demand of the oil field, the pier bases are settled on the narrow brick platform.

The vibration of the relief pipe may cause the pier bases to fall from the platform or may cause the platform to collapse, which in turn leads to the fixing becoming ineffective. In Figure 6-1-28, the installation and removal of the relief pipe is very inconvenient.

FIG. 6-1-27 Relief pipe is through a mud waste pit (A).   **FIG. 6-1-28** Relief pipe is through a mud waste pit (B).

Remedy

In order to connect, fix, dismantle, or install pipe, the relief pipe should not go through the mud waste pit, and should be laid along the ground instead (Figs. 6-1-29, 6-1-30). The relative position of the relief pipe and waste pit should be well considered before the implementation of the pre-spudding project. It is better not to put the relief pipe above the waste pit.

FIG. 6-1-29   Relief pipe is laid along the edge of a waste pit (A).

FIG. 6-1-30   Relief pipe is laid along the edge of a waste pit (B).

Hidden danger: The target side of the 90° buffer bent sub filled with lead faces the relief pipe outlet

Hazard

The target side of the 90° buffer bent sub (also called target-type sub, or T-bent sub) filled with lead faces the relief pipe outlet. The sub doesn't have erosion-resistant performance (Fig. 6-1-31).

FIG. 6-1-31   90° buffer bent sub is filled with lead in an incorrect installation.

FIG. 6-1-32   90° buffer bent sub is filled with lead in a correct installation.

Remedy

The target side of the 90° buffer bent sub filled with lead should face the relief pipe inlet (Fig. 6-1-32).

2. FIXING THE RELIEF PIPE

Hidden danger: The fixing point distance is over 15 m

Hazard

If the fixing point distance exceeds 15 m, it will make the pipelines unstable (Fig. 6-2-1). Also the vibration of the relief pipe will damage the air tightness of the pipelines, and can make the pipelines swing and possibly injure people who are rushing to deal with the emergency.

FIG. 6-2-1   Fixing distance of the relief pipe is about 18 m.   **FIG. 6-2-2**   Fixing distance of the relief pipe is 10 m.

Remedy

If the pier base is heavier than 1000 kg, the fixing point distance of the relief pipe should not exceed 15 m. When the weight of pier base is less than 1000 kg, the fixing point distance should be within 10 to 15 m (Fig. 6-2-2).

Hidden danger: The turnings of the relief pipe are not fixed or are fixed too far from the bent sub

Hazard

The turning of the relief pipe undergoes the greatest force when releasing. Therefore, if the turning points are not fixed or the fixing points are too far from the bent sub, the relief pipe will be quite unstable (Figs. 6-2-3, 6-2-4).

Remedy

The turning points of the relief pipe should be fixed firmly (Fig. 6-2-5), and the fixing point should be as close to the bent sub as possible (Fig. 6-2-6). Generally the distance between the fixing point and bent sub should be less than 1 m. The bent subs placed horizontally should be fixed in front and behind (Figs. 6-2-7, 6-2-8).

FIG. 6-2-3   Turning not fixed.

FIG. 6-2-4   Fixing point is about 3 m from the turning point.

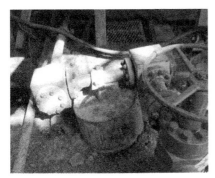

FIG. 6-2-5   Turning is fixed.

FIG. 6-2-6   Fixing point is near the bent sub.

FIG. 6-2-7   Turning is fixed in front and behind (A).

FIG. 6-2-8   Turning is fixed in front and behind (B).

Hidden danger: The fixing of the relief pipe is not firm enough

Hazard

If the fixing of the relief pipe is not firm enough, the vibration will damage the air tightness of the pipelines when releasing, and can even cause the pipelines to swing and possibly injure people who are rushing to deal with the emergency.

This can happen, for example, when the gland is not pressed tightly, the gland is too thin, the bolts are not screwed on tightly, a wooden block is under the gland, the foundation of the pier base is not solid, or material is leaking from the fill-in pier base (Figs. 6-2-9 to 6-2-14).

FIG. 6-2-9 Gland is not close to the pipe, and it's too thin.

FIG. 6-2-10 Gland is not fastened enough.

FIG. 6-2-11 Gland is not pressed enough.

FIG. 6-2-12 Foundation of the pier base is not solid.

FIG. 6-2-13 Fill-in pier base stuffing leaks.

FIG. 6-2-14 Fill-in pier base lacks stuffing.

Remedy 1

The relief pipe should be stable (Figs. 6-2-15, 6-2-16). It is better to use a whole-type gland, with a thickness and width of 5 mm and 100 mm, respectively. The diameter of bolts should be over 20 mm, and the screw nut should be installed firmly in place.

FIG. 6-2-15 Firmly fixed with fill-in pier base.

FIG. 6-2-16 Firmly fixed with cement pregrouted.

Remedy 2

Rubber can be used under the fixing gland of the relief pipe (mud recovery line). This will be an effective buffer (Figs. 6-2-17, 6-2-18).

FIG. 6-2-17 Rubber is filled under the gland of the relief pipe.

FIG. 6-2-18 Rubber is filled under the gland of the mud recovery line.

Hidden danger: Material leaks from the fill-in pier base

Hazard

Compared to the precast concrete pier base, the fill-in pier base is lighter and moves more easily. A poor design of the pier base structure or incomplete pier base components will cause material to leak from the pier base. Consequently, the pier base will not be heavy enough, which will affect the stability of the relief pipe (Figs. 6-2-19, 6-2-20).

FIG. 6-2-19 No baffle on the side door, and   **FIG. 6-2-20** Side door baffle does not seal.
the gland is on the surface.

Remedy

The fill-in pier base should not be designed with a door at the side, because the filling material can easily leak from a side door. The sand filling port should be set in the top of the pier base, which will prevent material from leaking, as shown in Figure 6-2-21. Figure 6-2-22 indicates another type of pier base filling called toy bricks cement. The fabrication and transportation of this type of pier base is very convenient and easy.

FIG. 6-2-21 Upper-port pier base.   **FIG. 6-2-22**   On bricks cement pier base.

It is better not to use the precast concrete pier bases because they are too heavy and difficult to transport, especially in mountainous areas, desert, marsh areas, and farmlands, where the hoist cannot be used. It is impossible to transport the 1000 kg bases in other ways.

Hidden dangers: The exit port fixed point of the relief line is too far from the exit; the exit port of the pipeline uses just one fixed footing/pier base

Hazard

The amplitude of the relief line's exit port/outlet is the largest when flow opens; if the fixed point is far from the exit, the stability will be poor. If there is only one fixed footing used on the exit port, the weight will be too light (Figs. 6-2-23, 6-2-24).

FIG. 6-2-23   The distance between the pier base and exit is over 4 m.

FIG. 6-2-24   There is only one footing used on the exit port.

Remedy

The industrial standard does not define an exact distance between the fixed footing and relief line's exit port. For ease of installation, the distance should not be over 1.5 m. According to the different pressures under the shaft, we can use two or more footings to fix the relief line's exit port, so that shaking will be lessened (Figs. 6-2-25 to 6-2-28).

FIG. 6-2-25   The distance between two footings/pier bases and the exit port of the pipeline is about 1 m.

FIG. 6-2-26   Use two footings to fix the exit port.

FIG. 6-2-27   Use a double footing to fix the exit port.

FIG. 6-2-28   Fix two parts of one footing/pier base.

If the weight of the exit port footing (ground anchor) is between 1600 and 2500 kg or greater, we can also fix two parts of one footing, just as Figure 6-2-28 shows.

Defect: The fixed footing is laid on the ground

Hazard

The weight of prefabricated footings (filled footing and cement footing) used so far is about 600 to 1000 kg, which is less than the weight of the industrial standard (1600–2500 kg). Putting the prefabricated footing on the ground will reduce the pipeline's stability (Figs. 6-2-29, 6-2-30).

FIG. 6-2-29   Filled footing is laid on the ground.

FIG. 6-2-30   Cement footing is laid on the ground.

Remedy

If the weight of a prefabricated footing is 1600 to 2500 kg, it is hard to move. In order to increase the stability of the pipeline and decrease the weight of the footing, we can embed the footing underground if the soil is easy to dig (Figs. 6-2-31, 6-2-32), especially for exploratory wells and wells with strata pressure over 35 Pa.

FIG. 6-2-31　Embed the footing underground (A).

FIG. 6-2-32　Embed the footing underground (B).

Defect: A welded fixed plate/gland is used

Hazard

The welded fixed plate/gland has a welded joint on the plate; as a result, it does not pass the welding quality test and there may be a welding quality defect (Figs. 6-2-33, 6-2-34). If the relief pipeline shakes, the welded joint may weaken, which will lead to the fixing losing efficacy.

FIG. 6-2-33　Welded fixed plate (A).

FIG. 6-2-34　Welded fixed plate (B).

Remedy

Use the integrative fixed plate (Figs. 6-2-35, 6-2-36).

FIG. 6-2-35   Integrative fixed plate (A).   **FIG. 6-2-36**   Integrative fixed plate (B).

Chapter 7

Common Hidden Dangers and Remedies of Inside Blowout Preventers (IBOPs)

Common hidden dangers from cocks:

- Cocks are often closed or opened halfway.
- There are no on or off marks or marks for the open hole position, or the marks are vague.
- The specifications of the spanner are not consistent.
- The handle of the cocks blocks the water also nozzle course; water channels are hindered.

Common hidden dangers of the drill string check valve:

- The drill string check valve lacks the top opening device.
- The water course of the open device is blocked.

Other hidden dangers:

- There is no drill pipe box thread on the top of the drill collar's lift sub.
- A BOP single is located in the mouse hole or on the slope ramp before the operation of trip with drill collar, the single is placed.
- The rig substructure height is greater than the 9 m used by the regular single blowout preventer.

1. COCKS

Hidden danger: The cock is often closed or opened halfway

Hazard

When a blowout happens in the drill string, the cock is often closed, and choking pressure is difficult (Fig. 7-1-1). If the cock is opened halfway, rapid flow in the well will wash out the spherical on-off device, invalidating the cock (Fig. 7-1-2).

Remedy

The cock should be opened during the awaiting orders operating mode (Fig. 7-1-3). The spherical on-off device should be fully opened to avoid being opened only halfway (Fig. 7-1-4).

FIG. 7-1-1   Cock closed.

FIG. 7-1-2   Cock half open.

FIG. 7-1-3   Cock open.

FIG. 7-1-4   Cock spherical on-off device open.

Hidden danger: There are no marks for on-off or for the open hole position, or the marks are vague

Hazard

There are no marks for on-off or for the open hole position, or the marks are vague. Whether the cock is open or not cannot be identified when attached to the drilling string (Figs. 7-1-5, 7-1-6).

FIG. 7-1-5   There is no on-off mark or mark for the open hole position.

FIG. 7-1-6   Marks for on-off or for the open hole position are vague.

Remedy

In order to grasp the cock's on-off state, there should be clear on-off marks and marks for the open hole position (Figs. 7-1-7, 7-1-8).

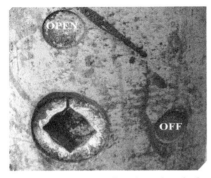

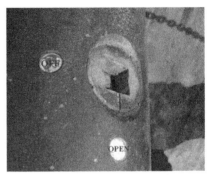

FIG. 7-1-7   Clear on-off marks and marks for the open or close position.

FIG. 7-1-8   Clear on-off marks and marks for the open hole position.

Hidden danger: The specification of the cock spanner is not consistent

Hazard

I have seen as many as five wrenches on the derrick floor. The wrenches are square or hexagonal inside, but are not the same size on opposite sides.

Cock wrench size is not uniform, so people often take the wrong wrench, even if the corresponding wrench appeared in the case.

When a blowout occurs and everyone is busy, it is easy to make mistakes. If the cock wrench cannot be found after the cock is connected, it would delay the time it takes to close the cock, and may even lead to an out-of-control blowout in the drill string (Figs. 7-1-9, 7-1-10).

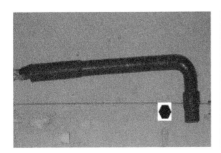

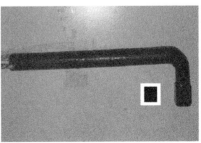

FIG. 7-1-9   Hex wrench.

FIG. 7-1-10   Quad wrench.

FIG. 7-1-11   The two different shapes of cock wrenches.

Remedy

The wrench standard for the Kelly lower cock, the BOP single upper cock, and the spare cock on the derrick floor should be unified. Because the Kelly upper cock has a larger outside diameter, daily use is less. Under normal circumstances, it is used only when the Kelly lower cock fails or the annular BOP seals the Kelly, so it is acceptable for the spanner and the cock not to be in the same specification.

Hidden danger: The cock handle blocks the water course

Hazard

To facilitate the handling or loading and unloading of cocks, cocks need to be installed with a handle. Currently handles come in a wide variety of forms. Some handles plug or block the water course (Figs. 7-1-12, 7-1-13). When trying to connect a cock, holding back pressure, especially when there is potential for a larger spray, makes it difficult to tighten.

FIG. 7-1-12   Cock water course is blocked (A).   **FIG. 7-1-13**   Cock water course is blocked (B).

Remedy

Hands should not block the water course, but should make the course an open road ahead (Figs. 7-1-14 to 7-1-17).

FIG. 7-1-14   Cock water course is unblocked (A).

FIG. 7-1-15   Cock water course is unblocked (B).

FIG. 7-1-16   Cock water course is unblocked (C).

FIG. 7-1-17   Cock water course is unblocked (D).

2.   DRILL STRING CHECK VALVE

Hidden danger: The drill string check valve lacks an open device

Hazard

When making a round trip, if the drilling device to open the top valve is missing, and you either grab the check valve or the check valve of the blowout preventer single, the water course will be blocked, which makes it difficult to connect the thread in a hurry. If the gushing is too strong, there is no way to connect the thread (Figs. 7-2-1, 7-2-2).

FIG. 7-2-1 BOP single doesn't have a check valve open device (A).

FIG. 7-2-2 BOP single doesn't have a check valve open device (B).

Remedy 1

The check valve on the derrick floor awaits orders, and the opening device of the check valve should be pre-installed, so that the water course would in-flow at any time. When making a round trip with a drill pipe, it is easy to rush-connect the thread.

When making a round trip with a drill collar, the check valve of the BOP single should be installed with an opening device; it is normally opened, which makes it easy to grab and connect to the drill collar. The handle of the check valve opening device will prevent hanging link rings. It is inconvenient to rush-connect a BOP single, as shown in Figure 7-2-3. To facilitate hanging link rings, a removable handle can be designed. After the opening device is connected to the BOP single, remove the handle (handle is connected to the opening device body with thread), as shown in Figure 7-2-4.

FIG. 7-2-3 Nonremovable handle makes it hard to hang link rings.

FIG. 7-2-4 Removable handle check valve opening device.

Remedy 2

While making a round trip with a drill pipe, the cock on the derrick floor should be open. While making a round trip with a drill collar, the cock should be connected to the BOP single ahead of time, and be open (Figs. 7-2-5, 7-2-6).

FIG. 7-2-5 BOP single connected with an open cock (A). **FIG. 7-2-6** BOP single connected with an open cock (B).

Hidden danger: The water course of the check valve opening device is blocked

Hazard

The opening device water courses of some check valves are blocked. It is unfavorable to rush-connect thread. In Figures 7-2-7 and 7-2-8, the channel of the opening device is opened at the side, and the blow out fluids of the drill string are changed into a flowing direction. The check valve is applied with an upward force. It is not easy to rush-connect thread from the side of the valve, where the gas relief fluid is not conducive to taking action to grab the wellhead staff.

Remedy

In order to rush-connect the thread conveniently, the water course of the opening device should be vertical and upward (Figs. 7-2-9, 7-2-10).

FIG. 7-2-7 The flow direction of fluids is changed by an opening device (A).

FIG. 7-2-8 The flow direction of fluids is changed by an opening device (B).

FIG. 7-2-9 Water course is vertical and upward (A).

FIG. 7-2-10 Water course is vertical and upward (B).

3.   OTHER HIDDEN DANGERS

Defect: The top of the drill collar lift sub has no box thread

Hazard

While drilling in a large borehole (borehole diameter more than 245 mm), generally, use two or three different sizes of drill collars. Some drilling crews are equipped with only one single BOP (Fig. 7-3-1). The crossover sub connected to the pin sub of the single BOP is only connected to the highest drill collar. Connect the single BOP when rising to the lower part of large-size drill collars, followed by a crossover sub (Fig. 7-3-2). For other well teams, with several drill collars and a few well sites on a single BOP, the BOP is connected under a single button type with different connectors.

Compared the second way with only a single blowout preventer, decreasing connection with the joint program, but a larger number of single BOPs, in emergency, it may pull a single case of wrong BOP. In addition, if the BOP is more than a single, routine maintenance workload increases slightly. Comparing the two methods, having more than one BOP single is more conducive to rapid shut-in to equipment.

If either one BOP single is equipped or many BOP singles are equipped, when making a round trip with a drill collar, the lift sub should be removed first, and then the single BOP is connected, so that shut-in time is longer. In some cases, completing the job within 5 minutes is difficult.

FIG. 7-3-1   Configuring one BOP single.

FIG. 7-3-2   Configuring two BOP singles.

Remedy

When making a operation of trip with a drill collar with a lift sub whose upper part doesn't have a box thread (Figs. 7-3-3, 7-3-4), the shut-in programs are as follows:

1. Remove the lift sub.
2. Connect the crossover sub.
3. Connect the BOP single.
4. Close the pipe-ram BOP after lowering the drill string.

Generally you need to go through the four steps before you can control the wellhead. Of course, if the crossover sub is connected to the BOP single ahead of time, you can eliminate step 3, but must follow the other three steps.

FIG. 7-3-3   Lift sub has no box thread (A).

FIG. 7-3-4   Lift sub has no box thread (B).

If the top of the lift sub has a box thread that can be connected to the drill pipe directly, the lift sub can be run in the hole. To make a round trip, the steps are as follows: (1) connect the BOP single; (2) close the pipe-ram BOP after lowering the drill string. Only two procedures can be completed on the well-head control. Compared to using a non-drill-pipe thread lift sub, reducing the removing lift sub, and connecting the crossover sub, these two steps help reduce the shut-in time (1 to 2 min), and make the operation simple and convenient.

If the upper part with drill pipe box thread of the lift sub is used, you just need to have one BOP single with an IBOP at the top of it. The lower part of the BOP single does not require any connection with crossover subs. Currently the lift sub with the drill pipe box thread improved penetration is not high, and some oil fields do not realize the positive significance of the tool. The use of this lift sub with a drill box thread should be vigorously promoted (Figs. 7-3-5, 7-3-6).

FIG. 7-3-5 Lift sub with a drill pipe box thread (A).

FIG. 7-3-6 Lift sub with a drill pipe box thread (B).

Defect: To avoid making a round trip with the drill collar, the BOP single is placed into a mouse hole or leaned against the ramp

Hazard

The preliminary BOP single purpose is to exchange and cut off the inside diameter of the drill string rapidly, when an overflow or blowout is discovered during a round trip with a drill collar, reducing shutting-in time. While making a round trip with a drill pipe, or while drilling, and you observe overflow or blowout shutting-in, there is no need to change the inside diameter of the drill string. The BOP single is placed into a mouse hole or leaned against the ramp. It interferes with employee operation, the operation security environment is deteriorated, and there is risk of injury to employees (Figs. 7-3-7 to 7-3-10).

FIG. 7-3-7   BOP single leaned against a ramp during drilling.

FIG. 7-3-8   BOP single in a mouse hole during a round trip with a drill pipe.

FIG. 7-3-9   BOP single on a ramp opening handle.

FIG. 7-3-10   BOP single leaned against a ramp during a round trip with a drill pipe.

Remedy

When making a round trip with a drill collar, the BOP single should be placed in a spot where the single can be reached and used easily. The spot can be a mouse hole or ramp (Figs. 7-3-11, 7-3-12). Under other conditions, the BOP single should be placed on the drill string racks.

FIG. 7-3-11   BOP single placed on the ramp (A).

FIG. 7-3-12   BOP single placed on the ramp (B).

Defect: The rig substructure height is more than 9 m while using a conventional BOP single

Hazard

The length of the drill pipe is generally 9.65 m, and if the rig substructure is over 9 m, the BOP ram may be located just below the drill pipe joints or the joints on the drill collar, and if the drill string and ram do not coincide with the diameter, the team cannot shut-in (Fig. 7-3-13).

FIG. 7-3-13   Height of the derrick floor is over 9 m.

FIG. 7-3-14   BOP single + short drill pipe + cock.

Remedy

The purpose of setting the BOP single is to rapidly convert the inside of the drill string to prevent a blowout or gush out of the inside hole of the drill string and to meet the size requirements of the ram. For the high substructure rig, a common BOP single cannot meet diameter size converted requirements. We must consider other alternatives. Using a BOP double-single, 19 m long, to make a round trip is very inconvenient. Another good choice used in some oil fields would be to connect a 9.65-m-long drill pipe with one short drill pipe that is 2 to 3 m (Fig. 7-3-14), and others use the 12-m-long drill pipe.

For a drilling rig equipped with a top drive drilling system (TDDS), it is more convenient to use a BOP stand. Whether to use a BOP stand should be based on each team's operating practices, as long as they reduce the shut-in time; choosing a BOP single or stand is acceptable.

Chapter 8

Liquid-Gas Separator Common Hidden Dangers and Remedies

Because there are no installation standards for liquid-gas separators, and because they have many outlets and inlets (liquid inlet, liquid outlet, gas outlet, drain outlet, and relief valve outlet), there are many hidden dangers on site.

Relief/safety valve common hidden dangers:

- The liquid-gas separator has no safety valve.
- The safety valve is connected with a slender and curved discharge tube.
- The safety valve relief port points to the wellhead or tank area.

Common hidden dangers of pressure gauge installation:

- The pressure gauge installed at the top of the separator is not inconvenient to observe.
- The pressure gauge is installed in the liquid-gas separator feed tube or container.

Liquid-feed pipe common hidden dangers:

- A welded rigid liquid-feed pipe is used on site.
- The liquid-feed pipe is connected with a union.
- An asbestos sealing ring flange is used in the liquid-feed pipe.
- The liquid-feed pipe is not fixed.

Liquid-discharging pipe common hidden dangers:

- A hand-disc valve is used to control the separator level.
- The outlet is lower than the inlet of the discharge pipe.
- The separator liquid (fluid outlet) is more than half the height of the separator tube.
- The outlet of the liquid-discharging pipe is installed on the tapered tank.
- The fluid outlet is near the bottom of the buffer tank.
- The liquid-discharging pipe uses rigid pipe.

Common hidden dangers of the gas relief pipe:

- The gas relief pipe diameter is less than the design diameter.
- The gas relief outlet is too close to the logging housing.
- The gas relief pipe emits gas at the wellhead.

Common hidden dangers of fixing separators:

- The separator is not fixed or not firmly fixed.
- The fixed line is not basket thread.
- Rope clamps are not used.
- The number of rope clamps is less than three, the rope clamps are reversely fixed, or the separator and guy lines are not anchored.

1. SAFETY VALVE

Hidden danger: The liquid-gas separator has no safety valve

Hazard

The liquid-gas separator is a pressure vessel, there is no safety valve, and with overpressure in the container, it cannot be protected.

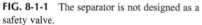

FIG. 8-1-1 The separator is not designed as a safety valve.

FIG. 8-1-2 The safety valves of the inlet tubes are sealed blind.

Remedy

Liquid-gas separators are pressure vessels, and should meet *Pressure Vessel Safety Technology Supervision Regulations TSG R0004-2009* and *GB 150-1998* requirements for overpressure relief devices (safety valve). The pressure relief valve automatically opens in the default, to prevent container damage due to overpressure and to protect personnel and equipment

(Figs. 8-1-3, 8-1-4). Valve models should meet the manufacturer's requirements; check it once a year.

FIG. 8-1-3   Installed with a safety valve.   FIG. 8-1-4   Safety valves installed.

Hidden danger: The safety valve is connected with a slender and curved discharge tube

Hazard

When the liquid-gas separator pressure exceeds the set pressure of the relief valve (safety valve set off artificial pressure), the gas discharge through the safety valve's slender curved tube will lead to choked flow, which will introduce back pressure. The back pressure prevents the safety valve from releasing smoothly, and if the pressure cannot be released quickly, it can result in over-pressure in the container. For ordinary spring-type safety valves, the back pressure over the set pressure must not exceed 10%.

In Figures 8-1-5 and 8-1-6, the use of hose as a discharge pipe may cause a sudden relief and the hose line to swing forcefully and injure nearby personnel. In Figures 8-1-6 and 8-1-8, there is a lot of electrical equipment in the fluid discharge outlet in the tank and the fuel gas discharge pipe gas relief-driven irrigation; combustible gas accumulation in the tank will explode. In Figures 8-1-7 and 8-1-8, in the gas relief pipe, there are two turns. The safety valve discharge pipe diameter is less than the outlet diameter, and the formation of cutting results in back pressure, which is not conducive to timely and quick relief. In Figures 8-1-9 and 8-1-10, the safety valve discharge pipe does not lead to the atmosphere. Connected with the gas relief pipe, the safety valve discharge pipe back pressure and tank pressure is almost the same. When the container is overpressured, more than the safety valve's set pressure due to back pressure with the same pressure vessel, the safety valve will not jump, and cannot protect the container pressure relief.

FIG. 8-1-5 Drain hose used in the safety valve.

FIG. 8-1-6 Safety discharge pipe located on the mud tank.

FIG. 8-1-7 Relief pipe reducing the diameter and bending (A).

FIG. 8-1-8 Relief pipe reducing the diameter and bending (B).

FIG. 8-1-9   Discharge pipe and relief pipe are linked through (A).

FIG. 8-1-10   Discharge pipe and relief pipe are linked through (B).

Remedy

The liquid-gas separator is installed on top of the guide valve. This valve does not allow the use of any back pressure, which allows the safety valve overflow of the gas surrounding the spread, not far from where the discharge pipe is connected. The liquid-gas separator safety valve on the discharge pipe can then lead to air emissions of a bent sub or a short tube leading to atmospheric emissions to meet the requirements.

In Figure 8-1-11, the best form for the discharge pipe uses a short tube for direct emissions.

In Figure 8-1-12, the discharge elbow exports up, but it should be drilled lower in the elbow. Emissions of rain, snow, dust, or other obstructions may be concentrated in the elbow, to avoid poor flow. In Figure 8-1-14 it is acceptable not to receive the discharge pipe.

FIG. 8-1-11   Direct discharge short tube.

FIG. 8-1-12   Discharge bent sub outlet up.

FIG. 8-1-13   Discharge elbow pipe down.

FIG. 8-1-14   Not connected with the discharge pipe.

Hidden danger: The safety valve discharge points to the wellhead or tank area

Hazard

The safety valve outlet points to the wellhead valve or tank area; when the liquid-gas separator is overpressured and the valve is off, combustible gas sprays to the wellhead and tank area. Because there are many combustible gas-driven electrical and other types of equipment, and the wellhead and tank, it is very dangerous (Figs. 8-1-15, 8-1-16).

FIG. 8-1-15   Outlet points to the tank area.

FIG. 8-1-16   Outlet points to the wellhead.

Remedy

Safety valves cannot discharge toward the equipment, platforms, ladders, cables, and such. The best discharge point is to the right side. If the device cannot be pointed to the right, then pointing to the front of the well site is an acceptable option (Figs. 8-1-17, 8-1-18).

FIG. 8-1-17   Outlet pointing to the right of the well site.

FIG. 8-1-18   Outlet pointing to the front of the well site.

2.   PRESSURE GAUGE

Defect: The pressure gauge installed at the top of the separator is not convenient to observe

Hazard

Most of the liquid-gas separator tank pressure gauges are mounted on the top of the gas part, which meets the *SY/T 0515-2007* requirements. But if the pressure gauge is installed in the liquid-gas separator at the top, from ground level over 7 m, it is not easy to observe (Figs. 8-2-1, 8-2-2).

Remedy

According to the manufacturer, the standard gauge should be installed in the gas separator at the top position. It is not easy to observe the pressure gauge when installed on the top, however, so it should be installed where it can be easily observed.

One approach is from the gas separator at the top position of pressure taps, cited 10 mm diameter seamless steel tubes under a pressure as a guide tube, at about 1.7 m from the ground to install a pressure gauge. This will not only locate the pressure gauge at the gas phase of the container, but also reduce the installation height to facilitate observation, as shown in Figure 8-2-3. The top of the pressure conductivity tube is connected with a tee joint, beside is connected with a pressure gauge and below is connected with a valve, which can link to pipeline to clean and empty the pressure conductivity tube.

FIG. 8-2-1 Pressure gauge is installed at the top of the tank (A).

FIG. 8-2-2 Pressure gauge is installed at the top of the tank (B).

Another approach is to install the gauge in the gas relief pipe on the separator, about 1.70 m from the ground, as shown in Figure 8-2-4. Although slightly lower than the pressure vessel, the difference is small and can be ignored. There are some well teams that take this approach.

FIG. 8-2-3 Pressure gauge is installed on the pressure-converted tube, and can be observed easily.

FIG. 8-2-4 Pressure gauge is installed on the gas relief pipe.

Hidden danger: The pressure gauge is installed in the liquid-gas separator pipe

Hazard

The liquid-gas separator liquid-feed pipe pressures are higher than that of the inside vessel; if the pressure gauge is installed in the inlet pipe, the displayed value will be too large and inaccurate (Figs. 8-2-5, 8-2-6).

FIG. 8-2-5   Pressure gauge is installed on the inlet buffer tube.

FIG. 8-2-6   Pressure gauge is installed on the liquid-feed pipe.

3.   LIQUID-FEED PIPE

Hidden danger: The liquid-feed hard pipe is welded on site

Hazard

The liquid-feed pipe or incoming stream manifold of the Liquid-gas separator can bear certain high pressure, if being welded on field, it is difficult to choose reasonable welding method and welding parameter, inspect the welding process and guarantee the welding quality.

FIG. 8-3-1   Liquid-in pipe is welded on site (A).

FIG. 8-3-2   Liquid-in pipe is welded on site (B).

Remedy

Do not allow on-site welding for cutting into the tube. Layout and connection is not convenient going into the tube using a hard line; if the liquid goes into the tube using a high-pressure fire hose, the layout and connection is a lot easier, and the pipelines will not need on-site welding (Figs. 8-3-3, 8-3-4).

FIG. 8-3-3 A joint of hose as the liquid-in pipe (A).

FIG. 8-3-4 A joint of hose as the liquid-in pipe (B).

Hidden danger: The liquid-in pipe is connected with a union

Hazard

The union buckle becomes loose by vibration, which makes it difficult to ensure the connection of the seal under pressure (Figs. 8-3-5, 8-3-6).

FIG. 8-3-5 Pipe connected with a union (A).

FIG. 8-3-6 Pipe connected with a union (B).

Remedy

For higher reliability, Into the tube, can withstand high-pressure pipeline steel ring flange seal (Figs. 8-3-7, 8-3-8). It is more safe to adopt flange with steel ring seal which can bear high pressure to connect with the the liquid-feed pipe.

FIG. 8-3-7   Using a steel ring seal flange (A).   FIG. 8-3-8   Using a steel ring seal flange (B).

Hidden danger: A flange with an asbestos seal ring is used in the liquid-in pipe

Hazard

The asbestos mat sealed flange can bear generally not more than 5 MPa, but the pressure of liquid-feed pipe is far more than 5 MPa, especially the inlet of the pipeline. It is dangerous to use the asbestos mat sealed flange, When choking and killing the well, the flange will leak because of stabbing, it is difficult to meet pipeline pressure requirements (Figs. 8-3-9, 8-3-10).

FIG. 8-3-9   Asbestos mat sealed flange (A).   FIG. 8-3-10   Asbestos mat sealed flange (B).

Remedy

The liquid-in pipe should be used to withstand the high-pressure pipeline steel ring flange seal (Figs. 8-3-11, 8-3-12).

FIG. 8-3-11 Connected with a steel ring seal flange (A).

FIG. 8-3-12 Connected with a steel ring seal flange (B).

Hidden danger: The inlet pipe is not fixed

Hazard

In the process of choking and killing, if the liquid-in pipe is not fixed or not firmly fixed, the vibration will affect the sealing performance of joints, affecting the valve sealing performance (Figs. 8-3-13, 8-3-14).

FIG. 8-3-13 Liquid-in hose not fixed.

FIG. 8-3-14 Liquid-in rigid pipe not fixed.

Remedy

The liquid-in pipe elbow and the middle line should have the pier base or anchor firmly fixed (Figs. 8-3-15, 8-3-16). The quality of the base pier should not be less than 650 k.

FIG. 8-3-15   Liquid-in hose firmly fixed (A).   **FIG. 8-3-16**   Liquid-in hose firmly fixed (B).

4.   LIQUID DIVERSION PIPE

Hidden danger: The height of the liquid level of the separator is used with a butterfly/disc valve

Hazard

If you rely on the manual control valve fluid control switch (Figs. 8-4-1, 8-4-2), the separator cannot be maintained in the normal work surface height. If you open the valve too long, it may make natural gas from the discharge port flow into the tank. If the open time is too short, the separator liquid level may be too high, the separator may be poor, and the gas relief pipe from the drilling fluid will overflow into the gas relief pipe.

FIG. 8-4-1   Manual control disc valve is installed to the outlet pipe (A).   **FIG. 8-4-2**   Manual control disc valve is installed to the outlet pipe (B).

Remedy

A liquid-gas separator with a level control system should be used to achieve the level of control, in the following two ways:

1. The use of a float ball to control the pneumatic valve to control the level, due to frequent failure of pneumatic valves, has rarely been used, as shown in Figures 8-4-3 and 8-4-5.
2. The use of a siphon control level, with simple structured, high reliability devices, is now widely used, as shown in Figures 8-4-4 and 8-4-6.

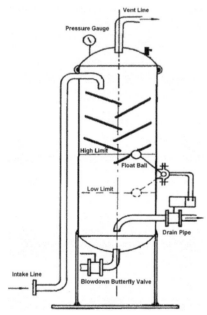

FIG. 8-4-3 Diagram of using a float ball to control the pneumatic valve to control the level.

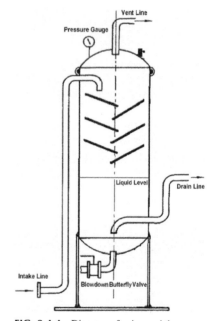

FIG. 8-4-4 Diagram of using a siphon control level.

FIG. 8-4-5 Use of a float ball to control the pneumatic valve to control the level.

FIG. 8-4-6 Use of a siphon control level.

Defect: The outlet is lower than the inlet of the liquid discharge pipe

Hazard

When the outlet is lower than the inlet of the liquid discharge pipe, the phenomenon in the siphon under the action of the separator above the discharge pipe will discharge the combustible gas separator liquid from the discharge pipe to the tank area (Figs. 8-4-7, 8-4-8).

FIG. 8-4-7 Outlet is lower than the inlet of the liquid discharge pipe (A).

FIG. 8-4-8 Outlet is lower than the inlet of the liquid discharge pipe (B).

Remedy 1

A fluid outlet discharge pipe should be higher than the entrance, the height of liquid in the separator should be a high level of 30%, or the minimum separator liquid and the liquid discharge port should be greater than 3 times the height difference between the discharge pipe diameter to avoid gas discharge through the discharge pipe (Figs. 8-4-9, 8-4-10).

FIG. 8-4-9 Outlet is higher than the inlet of the liquid discharge pipe (A).

FIG. 8-4-10 Outlet is higher than the inlet of the liquid discharge pipe (B).

Remedy 2

If the liquid separator outlet pipe is below the liquid inlet, a liquid pipeline can be set in the anti-siphon mouth. When the liquid level below the discharge pipe is in the highest position in the anti-siphon mouth of atmospheric pressure, the discharge pipe stops draining, to maintain a stable liquid level in order to avoid the gas going into the discharge pipe (Figs. 8-4-11, 8-4-12).

FIG. 8-4-11 Liquid discharge pipe being set with an anti-siphon mouth (A).

FIG. 8-4-12 Liquid discharge pipe being set with an anti-siphon mouth (B).

Hidden danger: The separator liquid (fluid outlet) is close to or more than half the height of the body of the separator

Hazard

The separator liquid level is too high, more than 30% of the cylinder's height, degassing is not sufficient, gas is inefficient, and the separator will reduce the amount of processing (Figs. 8-4-13, 8-4-14).

FIG. 8-4-13 Liquid level about 50% of the height of the cylinder.

FIG. 8-4-14 Liquid level about 60% of the height of the cylinder.

Remedy

For the gas efficiency and processing power to not be affected, the high level in the separator cylinder should be about 30%, which can handle the largest amount of degassing for the best results.

Take the discharge tube out from the bottom (Figs. 8-4-15, 8-4-16). This is equivalent to lengthening the cylinder height; in addition to gas efficiency, processing capacity is also increased. To prevent the gas from the tube from getting out, it is desirable to retain a certain level in the tank as a buffer. The height of liquid in the cylinder should be about 15 to 30%.

FIG. 8-4-15 Liquid discharge pipe taken out from the bottom of the tank (A).

FIG. 8-4-16 Liquid discharge pipe taken out from the bottom of the tank (B).

Defect: The pipe outlet is placed in the conical tank

Hazard

A liquid outlet is placed in the conical tank. Liquid from the gas relief gas outlet fluid, without a shaker except gas, is not conducive to further removing the drilling fluid in the gas (Figs. 8-4-17, 8-4-18).

FIG. 8-4-17 Pipe outlet is in the conical tank (A).

FIG. 8-4-18 Pipe outlet is in the conical tank (B).

Remedy

A liquid outlet is placed in the shaker at the entrance, through the separator discharge of drilling the second gas-shaker. The drilling fluid further reduces the air content (Figs. 8-4-19, 8-4-20).

FIG. 8-4-19 Pipe outlet is fixed in the inlet shale shaker (A).　　FIG. 8-4-20 Pipe outlet is fixed in the inlet shale shaker (B).

Hidden danger: There is a liquid outlet near the bottom of the buffer tank

Hazard

A liquid outlet is near the bottom of the buffer tank. The liquid outlet will be blocked from drilling cuttings in an elevated tank (Figs. 8-4-21, 8-4-22).

FIG. 8-4-21 A liquid outlet near the bottom of the buffer tank (A).　　FIG. 8-4-22 A liquid outlet near the bottom of the buffer tank (B).

Remedy

In order to prevent cuttings returned from the mud flume blocking the pipe outlet , the outlet should be installed in the upper section or at the top of the buffer tank (Figs. 8-4-23, 8-4-24).

FIG. 8-4-23   Outlet of pipe near the top of the buffer tank (A).

FIG. 8-4-24   Outlet of pipe near the top of the buffer tank (B).

Defect: A hard-line pipe is used in the discharge pipe

Hazard

The installation of a liquid outlet, located generally as conical or shaker cans at the entrance, with a hard line, is a bit inconvenient. The well and tube length are not the same, and the use of a hard piping installation is more convenient, so it cannot be treated as defective (Figs. 8-4-25, 8-4-26).

FIG. 8-4-25   Hard pipe serves as a discharge pipe (A).

FIG. 8-4-26   Hard pipe serves as a discharge pipe (B).

Remedy

The discharge pipe pressure is below the rated pressure the of liquid-gas separator. For easier installation, you can use the hose as a discharge pipe (Figs. 8-4-27, 8-4-28).

FIG. 8-4-27 Hose serves as a discharge pipe (A).   **FIG. 8-4-28** Hose serves as a discharge pipe (B).

5. GAS RELIEF PIPE

Hidden danger: The gas relief pipe diameter is less than the design diameter

Hazard

The gas relief pipe diameter is less than the design diameter, so the gas relief velocity will decrease, which may lead to separator overpressure and repeated take-off of the safety valve, so the separator does not work (Figs. 8-5-1, 8-5-2).

FIG. 8-5-1 Gas relief pipe diameter is less than the design diameter (A).   **FIG. 8-5-2** Gas relief pipe diameter is less than the design diameter (B).

Remedy

The diameter gas relief pipe should be designed to select the diameter. In the choice of the gas relief pipe diameter, you should take into account the ease of connection and pipe diameter. Pipe stiffness is also important, because the undulating landscape can be slightly difficult on the flange.

FIG. 8-5-3   Diameter of gas relief pipe is unified with the design diameter.

FIG. 8-5-4   Gas relief pipe diameter is little less than the design diameter.

Hidden danger: The gas relief outlet is too close to the logging housing

Hazard

The logging house is usually placed on the left of the field as shown in Figures 8-5-5 and 8-5-6. The distance between the flare line outlet and the logging house is just about 3–5 m, and igniting the inflammable gas at the outlet will burn the logging house.

FIG. 8-5-5   Gas relief pipe near the logging house (A).

FIG. 8-5-6   Gas relief pipe near the logging house (B).

Remedy

The gas relief pipe should be connected 50 m beyond the well site, with a vertical combustion tube. The logging room should be at least 20 m away from the gas relief outlet. The gas relief outlet should be away from the logging house to prevent it from igniting, as in Figures 8-5-7 and 8-5-8.

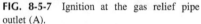

FIG. 8-5-7 Ignition at the gas relief pipe outlet (A).

FIG. 8-5-8 Ignition at the gas relief pipe outlet (B).

Hidden danger: The gas relief pipe emits gas at the wellhead

Hazard

The liquid-gas separator is generally about 12 to 16 m from the wellhead. If flammable gas is released directly on the spot, it will gather around the rig, making it easier to reach a lower explosive limit, which is dangerous. This discharge method cannot be implemented at the ignition exit. You cannot prevent the burning of combustible gases once they ignite (Fig. 8-5-9).

FIG. 8-5-9 Gas is released directly from the gas relief outlet on the spot.

Remedy

The gas relief outlet location can be chosen in two ways. One way is to pick a well site 50 m away (*SY/T 5964-2006* states that "liquid-gas separator diameter gas relief pipe connection according to design a well field 50 m beyond," and *SY/T 6426-2005* provides that "the liquid-gas separator connected the gas relief pipe out of the wellhead more than 50 m"; the two standard gas relief pipe length requirements are inconsistent), as shown in Figure 8-5-10. Another way is to connect the flare line along the derrick leg to the derrick crown. An artificial island in Jidong Oilfield and offshore drilling in China as well as some foreign surface oilfield adopt this approach (Fig. 8-5-11). Both methods discharge combustible gases at the exit, so as not to ignite a fire.

FIG. 8-5-10 Gas relief pipe is connected 50 m beyond the well site.

FIG. 8-5-11 Gas relief pipe is connected to a crown block.

6. SEPARATOR

Hidden danger: The separator is not fixed or not firmly fixed

Hazard

If the separator is not fixed or not firmly fixed, when in use the vibration will affect the tightness of the joints and cause separator shift. Severe vibration may cause the collapse of the separator (Figs. 8-6-1, 8-6-2).

Remedy

The separator should be fixed; you can use 16 or 12 mm square fixed rope, or you can use three guy wires separated by 120° (Figs. 8-6-3, 8-6-4).

FIG. 8-6-1   Separator is not fixed.

FIG. 8-6-2   Missing one piece of guy line.

FIG. 8-6-3   Fixed with four guy lines.

FIG. 8-6-4   Fixed with three guy lines.

Hidden danger: A fixed basket guy line screw is used

Hazard

When the guy line is loose, remove the rope clamps to adjust it if there is no steamboat ratchet/bolt (Figs. 8-6-5, 8-6-6). Disassembling the rope clamps to adjust the tightness of the rope is not only inconvenient, but it creates operational difficulties. When using the separator, it is impossible to adjust guy lines by removing the rope clamps.

FIG. 8-6-5   Guy line does not use a basket bolt (A).

FIG. 8-6-6   Guy line does not use a basket bolt (B).

Remedy

It is easiest to adjust guy line tightness with a steamboat ratchet (Figs. 8-6-7, 8-6-8).

FIG. 8-6-7   Basket bolt.

FIG. 8-6-8   Basket bolt is not used in a guy line.

Hidden danger: The guy line is not used with rope clamps, the number of rope clamps is less than three, or the rope clamps are fixed reversely

Hazard

The guy line/wire is not used with rope clamps, as shown in Figure 8-6-9. The number of rope clamps is less than three, they are not firmly fixed, or the line end may be detached as shown in Figure 8-6-10. The U-bolt of the rope clamp is stuck in the main line, and it will damage the wire rope, decreasing the strength of the guy line, as shown in Figures 8-6-11 and 8-6-12.

FIG. 8-6-9 Guy line does not use rope clamps.

FIG. 8-6-10 Two rope clamps are used.

FIG. 8-6-11 U-bolt is fastened to the main line.

FIG. 8-6-12 Top rope clamp is fixed reversely.

Remedy

The up and down sides of the guy line use three rope clamps. The clamp seat is stuck on the servicing interval of the guy line. The U-bolt is stuck in the tail section/interval of the line. The distance of the rope clamp is 150 to 200 mm. Correct methods are shown in Figures 8-6-13 and 8-6-14.

FIG. 8-6-13 Number of rope clamps, clamp distance, and manner are correct (A).

FIG. 8-6-14 Number of rope clamps, clamp distance, and manner are correct (B).

Hidden danger: The guy line and separator are not anchored

Hazard

The guy line and separator are not anchored and are not fixed firmly (Figs. 8-6-15, 8-6-16).

FIG. 8-6-15 Guy line and separator are not anchored (A).

FIG. 8-6-16 Guy line and separator are not anchored (B).

Remedy

The guy wire and separator hanger should be anchored, and three or four guy lines should be used to anchor them evenly (Figs. 8-6-17, 8-6-18).

FIG. 8-6-17 Guy line and separator are anchored (A).

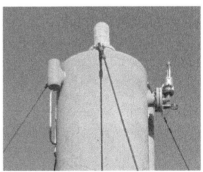

FIG. 8-6-18 Guy line and separator are anchored (B).

Chapter 9

Common Hidden Dangers and Remedies of Vacuum Degassers

Common hidden dangers in installing a vacuum degasser:

- The degasser is not connected with a flare line pipe, or the pipe is not at least 15 m from the mud tank area.
- The relief pipe of the degasser shrinks in diameter or is bent, so that the gas is not delivered smoothly.
- The relief pipe of the degasser is upright, releasing gas above the mud tank area.
- The degasser is installed on the suction tank of the mud pump.
- The inlet and outlet of the degasser are installed in one tank.
- The degasser is installed on the tank before the mud recovery line outlet.

1. VACUUM DEGASSER

Hidden danger: The degasser is not connected with a flare line pipe, or the relief pipe is not connected at least 15 m beyond the tank area

Hazard

The degasser is not installed with a relief pipe, or the pipe connected outside the tank is less than 15 m away. The gas discharged will accumulate in the tank area, causing danger of explosion (Figs. 9-1-1, 9-1-2).

FIG. 9-1-1   Without a relief pipe (A).   **FIG. 9-1-2**   Without a relief pipe (B).

Remedy

The relief pipe of the degasser is connected 15 m beyond the tank area (Fig. 9-1-3).

The flammable pernicious gases can be ignited at the wellhead if necessary. In Figure 9-1-4, the 15 m relief pipe is connected beyond 15 m when the degasser is used.

FIG. 9-1-3 Relief pipe is connected 15 m   **FIG. 9-1-4**   15 m relief pipe is connected.
beyond the tank area.

Hidden danger: Reduced diameter or a bent relief pipe of the degasser blocks the passage of gas

Hazard

If the relief pipe of the degasser has a reduced diameter or is bent, it decreases the efficiency of removing gases (Figs. 9-1-5 to 9-1-8).

FIG. 9-1-5   Bent relief hose (A).

FIG. 9-1-6   Bent relief hose (B).

FIG. 9-1-7   Relief hose is reduced in diameter at bend.

FIG. 9-1-8   Relief rigid pipe is reduced in diameter.

Remedy

The inside diameter of the relief pipe of the degasser should not be less than the open diameter of the outlet, and should open toward the road ahead. If the hose is used, it should not be deformed due to compression; a right angle bent sub should be used to turn a corner (Figs. 9-1-9, 9-1-10).

FIG. 9-1-9 Right angle is used at a corner turn (A).

FIG. 9-1-10 Right angle is used at a corner turn (B).

Hidden danger: The relief pipe is upright, discharging gases above the tank area

Hazard

The relief pipe discharges directly above the tank area. Because the outlet of the pipe is about 3 m above the surface of the tank, pernicious gases cannot be ignited and set on fire. Inflammable pernicious gases will accumulate around the tank area (Figs. 9-1-11, 9-1-12).

FIG. 9-1-11 Gas relief pipe releasing up (A).

FIG. 9-1-12 Gas relief pipe releasing up.

Remedy

The relief pipe of the degasser is connected 15 m beyond the tank area. Inflammable pernicious gases can burn at the outlet when released.

Hidden danger: The degasser is installed on the suction tank of the mud pump

Hazard

If the degasser is installed on the suction tank of the mud pump, there's not enough time for the gas-bearing mud circulated from the well to be degassed; it is sucked into the borehole by the mud pump. If the gas-bearing mud goes into the annulus, the hydraulic column pressure decreases further, and well invading will be sped up (Fig. 9-1-13).

Remedy

The installation position of the degasser should meet the gas-bearing mud returned from the borehole. First it is degassed by the degasser, and then it goes into the mud tank to be circulated. The degasser should be installed between the shale shaker and desander (Fig. 9-1-14).

FIG. 9-1-13 Degasser is near the suction tank of the mud pump.

FIG. 9-1-14 Degasser is between the shale shaker and desander.

Hidden danger: The inlet and outlet of the degasser are both in the same tank

Hazard

The inlet and outlet of the degasser are in the same tank. The mud degassed by the degasser returns to the tank, degassing repeatedly (Figs. 9-1-15, 9-1-16).

FIG. 9-1-15   Inlet and outlet of the pipe are in   **FIG. 9-1-16**   Inlet and outlet of the pipe are in
the same tank (A).   the same tank (B).

Remedy

The inlet and outlet of the degasser should not be in the same tank. The outlet is located in another tank under the inlet. The outlet is also located in the circulating ditch, but the outlet must be 200 to 350 mm under the liquid level. The inlet should be 300 to 550 mm from the bottom of the separating settling tank, and the tank storehouse should have a mixing function well. A filter should be installed at the inlet (Figs. 9-1-17, 9-1-18).

FIG. 9-1-17   Outlet is in the circulating ditch.   **FIG. 9-1-18**   Pipe outlet is in another tank.

Hidden danger: The degasser is installed on the tank before the mud recovery line outlet

Hazard

The degasser is installed on the tank before the mud recovery line outlet (Figs. 9-1-19, 9-1-20). During operation of the choking and killing well, the gas-bearing mud flowing from the mud recovery line does not go into the degasser. It circulates into the suction tank, pumping it into the borehole again. In this way, first the inflammable gases will accumulate on the surface of the tank. Once the gas concentration achieves the lower limit of explosion, it is likely to explode when it encounters sparks. Second, when the gas-bearing mud is pumped into the borehole, the hydraulic column pressure of mud will decrease. It is unfavorable to recover the pressure balance in the well, and it will cause the kill well operation to fail.

FIG. 9-1-19   Degasser is located before the recovery line outlet.

FIG. 9-1-20   Degasser is located behind the recovery line.

Remedy

The degasser is installed on the tank behind the mud recovery line outlet, or the degasser and the mud recovery line outlet are located in the same tank (Figs. 9-1-21, 9-1-22).

FIG. 9-1-21   The degasser is installed behind the mud recovery line outlet.

FIG. 9-1-22   Degasser and recovery line are in the same tank.

Chapter 10

Common Hidden Dangers and Remedies for Fireproofing, Explosion-proofing, and Preventing Hydrogen Sulfide

Common hidden dangers of fireproofing safety measures include:

- The fire extinguisher is in the open air outdoors.
- Different types of powder fire extinguishers are used together.
- BC types of powder fire extinguishers are equipped/supplied.
- The fire extinguisher exceeds its expiration date.
- Not enough fire extinguishers are supplied.
- Fire extinguishers are not inspected and maintained regularly.
- Carried by hand, the machine-operation firefighting pump is in the firefighting room.

Common hidden dangers of explosion-proofing include:

- There is a non-explosion-proof electric appliance within 30 m of the wellhead.
- The fixing bolts on the separated-explosion shell do not fit tightly.
- Explosion-proof appliances are used outside the explosion-proof area.
- There are no cooling and clearing spark devices.
- The gas relief pipe of the diesel engine cooling and spark eliminated devices increase power loss.
- The gas relief pipes of the diesel engine are toward the wellhead or circulating tank area.
- The boiler house is on the leeward.
- The distance between the oil tank and generating house is less than 20 m.

Common hidden dangers of safety measures to prevent hydrogen sulfide include:

- The hydrogen sulfide monitor is higher than the monitoring level.
- The non-explosion-proof alarming monitor of hydrogen sulfide is applied in the explosion-proof area.
- The current well site conditions are unclear from the warning sign.

- A warning sign is hung in a well site without hydrogen sulfide.
- Positive pressure respirators are in the facilities room.
- There is no respirator compressor.

1. SAFETY MEASURES OF FIREPROOFING

Hidden danger: Fire extinguishers are placed in the outdoor open air

Hazard

Not only in the scorching summer, but also in the northern cold winter, some drilling or operating crews put fire extinguishers directly in the outdoor open air, with no basking-proof or heat prevention measures taken (Figs. 10-1-1 to 10-1-4). The critical temperature of fire extinguishers is 55°C. If the temperature is exceeded, the fire extinguisher will explode because of high pressure. Also sunlight damages the rubber hose, which affects the spraying property, shortening the spraying range.

FIG. 10-1-1   Fire extinguisher in the open air.

FIG. 10-1-2   Fire extinguisher placed outdoors (less than –20°C).

FIG. 10-1-3   Fire extinguisher placed outdoors (A).

FIG. 10-1-4   Fire extinguisher placed outdoors (B).

The minimum temperature of powder fire extinguishers, carbon dioxide fire extinguishers, or methane halide fire extinguishers made in China is −20°C. If the temperature is less than that, the fire extinguisher will not spray extinguishant because of the lower temperature. The water type and foam fire extinguishers appeal higher circumstance temperature, so they are not applied to drilling, work-over and oil testing on site, a fat lot application at present.

Remedy

The circumstance temperature of the fire extinguisher must be within the range of the temperature of use (−20°C to 55°C), ensuring its function and security. The fire extinguisher should be kept indoors in winter, but far away from the room's heat source. The fire extinguisher should be not kept in the sun in summer; protective measures, such as keeping it in the box, should be taken to prevent the extinguisher from overheating (Figs. 10-1-5, 10-1-6).

FIG. 10-1-5   Taking protective measures.

FIG. 10-1-6   Fire extinguisher in the box (in summer).

Hidden danger: Different types of powder fire extinguishers are used at the same time

Hazard

Some drilling crews are equipped with two types of fire extinguishers, ABC and BC, which are used at the same time (Figs. 10-1-7, 10-1-8).

While firefighting, ABC and BC types of fire extinguishers can't be used at the same time. The two types of extinguishants will react to each other and greatly decrease the extinguishing function.

FIG. 10-1-7 ABC powder fire extinguisher.　　**FIG. 10-1-8** BC powder fire extinguisher.

Remedy

While firefighting, ABC and BC types of fire extinguishers can't be used at the same time. The two groups of fire extinguishers are known to be incompatible. Drilling crews should not be equipped with two types of incompatible fire extinguishers at the same time; they should be equipped with only one type of powder fire extinguisher. It is proposed that they be equipped with ABC powder fire extinguishers and phosphate ammonium extinguishers.

Defect: The crew is equipped only with BC powder fire extinguishers

Hazard

A crew may encounter solid, liquid, gas, and electric fires, but a BC powder fire extinguisher is suitable only for liquid, gas, and electric fires. It has no attaching function to a solid burning object, and is used only to control the fire—it can't extinguish it. So a BC powder fire extinguisher is not suitable for solid fire.

Remedy

Available fire extinguishers should be effective for solid, liquid, gas, and electric fires. ABC powder fire extinguishers can meet these requirements (Figs. 10-1-9, 10-1-10).

Hidden danger: The fire extinguisher exceeds its expiration date

Hazard

It is difficult to ensure a fire extinguisher's performance if it has expired, and it may cause an explosion (Figs. 10-1-11, 10-1-12).

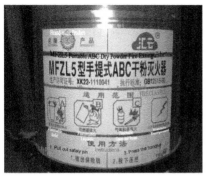

FIG. 10-1-9 ABC powder fire extinguisher (A).

FIG. 10-1-10 ABC powder fire extinguisher (B).

FIG. 10-1-11 Powder fire extinguisher explosion in a residential area.

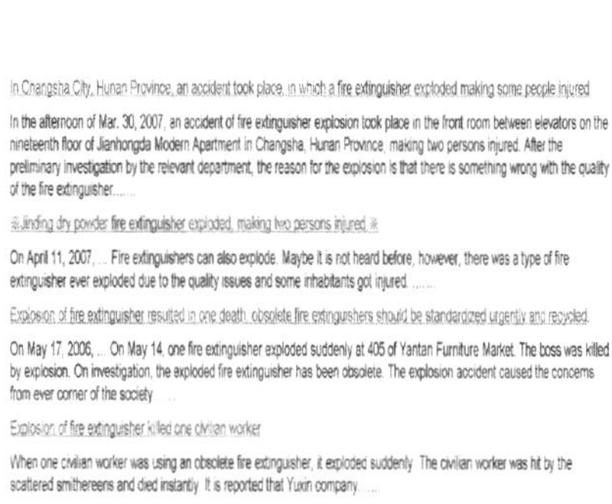

In Changsha City, Hunan Province, an accident took place, in which a fire extinguisher exploded making some people injured

In the afternoon of Mar. 30, 2007, an accident of fire extinguisher explosion took place in the front room between elevators on the nineteenth floor of Jianhongda Modern Apartment in Changsha, Hunan Province, making two persons injured. After the preliminary investigation by the relevant department, the reason for the explosion is that there is something wrong with the quality of the fire extinguisher....

Jiuding dry powder fire extinguisher exploded, making two persons injured, it

On April 11, 2007, ...Fire extinguishers can also explode. Maybe it is not heard before, however, there was a type of fire extinguisher ever exploded due to the quality issues and some inhabitants got injured. ......

Explosion of fire extinguisher resulted in one death, obsolete fire extinguishers should be standardized urgently and recycled

On May 17, 2006, ... On May 14, one fire extinguisher exploded suddenly at 405 of Yantan Furniture Market. The boss was killed by explosion. On investigation, the exploded fire extinguisher has been obsolete. The explosion accident caused the concerns from ever corner of the society ...

Explosion of fire extinguisher killed one civilian worker

When one civilian worker was using an obsolete fire extinguisher, it exploded suddenly. The civilian worker was hit by the scattered smithereens and died instantly. It is reported that Yixin company...

FIG. 10-1-12 The news that an expired fire extinguisher led to human death and injury.

Remedy

According to *Fire Extinguisher Maintenance and Abandon Regulations GA 95—2007*, counting from its release from manufacture, a fire extinguisher must be abandoned when it reaches its expiration limit as follows: powder fire extinguisher, 10 years; carbon dioxide fire extinguisher and stored gas vase, 12 years.

Hidden danger: The site is not equipped with enough fire extinguishers

Hazard

The number of fire extinguishers equipped is below the minimum numbers required by industry standard *SY 5974—2007*. It is difficult to meet extinguishing needs at this level.

Remedy

The number of fire extinguishers available shouldn't be under industry standard *SY 5974—2007* requirements: two 100 kg or six 35 kg extinguishers, ten 8 kg powder fire extinguishers, and two carbon dioxide fire extinguishers in the fire controlling room should be provided. The engine house should be equipped with three 8 kg carbon dioxide fire extinguishers. The electric house should be equipped with two 8 kg carbon dioxide fire extinguishers (Fig. 10-1-13).

FIG. 10-1-13   Extinguisher in the fire control-   **FIG. 10-1-14**   Powder fire extinguishers on ling room.   mud tanks.

SY 5974—2007 has regulated the equipment extinguisher to fire extinguisher in the fire controlling room, engine room, and electric house at the well site. Whether the regulations that fire extinguishers should be equipped on the surface of mud tanks, the derrick floor, in houses with electric devices, the BOP remote console, the oil tank area, and such have not been worked out. According to *Specifications of Architecture Extinguisher Collocated Designed GB 50140—2005*, the places just mentioned should also be equipped with fire extinguishers as follows: four 8 kg powder extinguishers for mud tanks (including the mud treating stuff desk; Fig. 10-1-14); two 8 kg powder extinguishers on the derrick floor; two 8 kg carbon dioxide extinguishers for the electric devices controlling house ; two 8 kg powder extinguishers in the remote console; and two 35 kg Cart-type powder extinguishers in the oil tank area.

Hidden danger: Extinguishers are not maintained and checked regularly

Hazard

Powder extinguishers and carbon dioxide extinguishers haven't been maintained since they came out of the factory. It is difficult to insure their performance, and it is likely to cause an explosion.

Regular inspections are not thorough, so defects aren't detected in time, which leads to an extinguisher that cannot be used, or causes an explosion during servicing. In Figure 10-1-15, the pressure indicator of the extinguisher is in the yellow area; it is likely to explode because of super pressure. In Figure 10-1-16, there are obstacles all around the extinguisher, making it difficult to reach when needed.

FIG. 10-1-15   Pressure indicator of the extinguisher is in the yellow area.

FIG. 10-1-16   Inability to fetch and use the extinguisher because of the obstacles all around it.

Remedy

According to *GB 50444—2008*, powder extinguishers and carbon dioxide extinguishers should be maintained 5 years after they come from the factory. After this first maintenance, they should be maintained every two years. The maintenance should be done by a specialized maintaining unit.

Extinguishers should be checked at least every month. For drilling or if the operating time is less than one month, inspection should be done on each well at least. The items to be inspected are as follows:

- Whether it meets maintenance conditions and time limit
- Whether there are rain-proof, sun-proof, etc. protective measures
- Whether there are obstacles around the extinguisher, preventing its use
- Whether the nameplate is toward the outside
- Whether the nameplate is damaged or clear
- Whether the lead seal, pin bar, etc. safety devices are damaged or lost (Fig. 10-1-18)
- Whether there are clear scathes (knock damage, lacerating damage), defects, or leakage on the barrel
- Whether the spouting hose is perfect; there is no patent chap, the nozzle is not blocked
- Whether driving gas pressure is in the range of service (whether the pressure indicator of a store-type extinguisher is in the green area (Fig. 10-1-17); a carbon dioxide extinguisher and storing gas bottle can be inspected with weighing method)
- Whether parts and components are complete, and there is no loosening, breaking off, or damaged marks

FIG. 10-1-17 Pressure indicator is in the green area.

FIG. 10-1-18 Lead seal and the safety pin are perfect.

Defect: The fire controlling pump is laid manually in the fire controlling house

Hazard

If the fire controlling pump is laid in the house (Fig. 10-1-19), during a fire, two men are needed to carry the pump (mass about 50–60 kg) at least to the water tank, and then connect the water supply pipe and discharge pipe. Then the diesel engine is started to make the pump run. The water gun spouts water. Generally, this course takes about 10 minutes. When the water gun spouts water, the fire has likely progressed from its beginning to its development, and possibly to burning violently.

FIG. 10-1-19 Fire controlling pump is manually set in the fire controlling house.

FIG. 10-1-20 Fire controlling pump is powered by a gasoline engine.

The pump is powered by a gasoline engine (Fig. 10-1-20), but a gasoline engine can't be operated by anyone; it requires personnel with specialized knowledge to operate. Laypeople are strictly forbidden to operate it, and is restricted to dealing with emergencies. Daily maintenance of the gasoline engine is also very bothersome and scrupulous work. If maintenance and use are not correct, it will affect the characteristics of the pump.

Remedy 1

Put the fire controlling pump manually beside the water tank. The discharging pipe and hose are connected ahead of time, so that it will shorten the preliminary time when starting the pump. But if the pump is set outdoors, some sun-proofing measures should be taken; otherwise, it is difficult to ensure the pump's performance and it will shorten the pump's life.

In Figure 10-1-21, the fire pump has been connected with a discharging pipe, but there are no sun-proofing measures. In Figure 10-1-22, the discharging pipe is connected ahead of time, and the pump is put in the protecting box, preventing it from sun damage and wind and sand invasion.

FIG. 10-1-21   Fire pump is connected with pipe.

FIG. 10-1-22   Fire pump is in the protecting box.

Remedy 2

A fixed electrodynamics firefighting pump is set at the water tank, replacing the move-by-hand pump with a machine-driven firefighting pump (Figs. 10-1-23, 10-1-24). Compared to a move-by-hand pump, a machine- driven firefighting pump's daily maintenance is simple and convenient, its operation is simple and swift, and a professional is not needed for operation. But note that the electric firefighting pump should be connected with a special electric line from the generating house. Once electricity from the electric distribution house is cut off, the electric firefighting pump will lose power.

FIG. 10-1-23   Fixed electric firefighting pump at the water tank (A).

FIG. 10-1-24   Fixed electric firefighting pump at the water tank (B).

Remedy 3

Install the firefighting pipe connector in the supplied water pipe (Figs. 10-1-25, 10-1-26). When a fire occurs, after the fire hose is connected, anyone can start the water pump and extinguish the fire. It is simple and convenient to maintain and operate.

FIG. 10-1-25 Install the firefighting pipe connector in the supplied water pipe (A).

FIG. 10-1-26 Install the firefighting pipe connector in the supplied water pipe (B).

Comparing the three methods, use of the fixed electric firefighting pump or installation of the firefighting pipe connector in the supplied water pipe is simple and practicable, and it is more suitable to an on-site drilling operation. Of course, if electricity is cut off during a fire, the move-by-hand, machine-driven pump has a definite advantage.

Remedy 4

Industry Standard SY/T 5225—2005 specifies only that "the fire fighting pipe connector should be installed in the supplied water pipe, and spare fire hose and water gun for operation of exploratory well, high pressure oil and gas well." So except for an exploratory well, or a high pressure oil and gas well, installation of the fire-fighting pipe connector in the supplied water pipe is certainly not an option.

2.   EXPLOSION-PROOF SAFETY MEASURES

Hidden danger: There are non-explosion-proof electric devices within 30 m of the wellhead

Hazard

SY/T 5225—2005 regulates that all electric devices within 30 m of the wellhead should have explosion-proof characteristics. If one of them is not explosion-proof in the area, the entire area will be non-explosion-proof. Once combustion gases get together and reach an explosive limit, it is likely to explode. Because explosion-proof knowledge training is limited, or crews don't think much of it, it is common for non-explosion-proof electric devices to be within 30 m of the wellhead (Figs. 10-2-1 to 10-2-6).

FIG. 10-2-1 Electric socket, six-speed visco-simeter, and electric torch in the duty house are not explosion-proof.

FIG. 10-2-2 The telephone in the driller's house is not explosion-proof.

FIG. 10-2-3 There are non-explosion-proof facilities in the tang man house because devices are within 15 m of the wellhead.

FIG. 10-2-4 The electric devices in the electric distribution cabinet on the mud tank are not explosion-proof.

FIG. 10-2-5 Electric devices in the dog house are not explosion-proof.

FIG. 10-2-6 The electric distribution cabinet under the derrick floor is not explosion-proof.

Remedy

The area within 30 m of the wellhead is a dangerous area. All electric devices in the area must be explosion-proof.

For lower pressure (below 21 MPa) switches, development wells, or heavy oil wells, the dangerous area should be reduced, and the area within 15 m from the wellhead is determined as a dangerous explosive site; that is, the electric devices within 15 m of the wellhead must be explosion-proof (Figs. 10-2-7, 10-2-8). Refer to *SY 6586—2003*.

FIG. 10-2-7 Explosion-proof electric devices in the dog house.

FIG. 10-2-8 Explosion-proof telephone, switches in the driller's house.

In some standards, there are some error explosion-proof requirements for drilling with electric devices. For example, *SY 6586—2003* regulates "well site, derrick floor, derrick, dog house, power and pump houses, electric devices of cleaning system, lightening apparatus, switches, buttons, electric cabinet should conform to explosion-proof requirements." It is not necessary that all electric devices in the well site be up to standard. And there is lack of relevance. If the derrick floor, derrick, dog house, power and pump houses, electric devices of the cleaning system, lightening apparatus, switches, buttons, and electric cabinet should conform to explosion-proof requirements, it can be done, and it is necessary. Other standards such as *AQ 2012—2007* and *SY/T 5974—2007* also have similar requirements, again with a lack of relevance. For example, the boiler, generator, computers, printers, etc. in the geological house have no explosion-proof performance. Even if the explosion-proof lights, switches, buttons, and such are used in these sites, it is out of the question that the explosion-proof requirement of all the equipment is being realized.

At present, explosion-proof lights, switches and buttons in the boiler house, geo-house, generating house, and cadre duty house of some drilling crews (Figs. 10-2-9, 10-2-10).

Logging unit houses as explosion-proof property can be divided into two types. One is a positive pressure explosion-proof logging unit house (Fig. 10-2-11). All doors and windows in the house are sealed. Making use of the blower more than 30 m from the wellhead, blow wind and pressurize into the house via pipe, raising the pressure in the house over one atmospheric pressure all the time. It is difficult for combustion gases that spread out from the well site to enter the house. Even if the pressure is lost, there is a detector. Trigger the alarm first, and then turn off the general electrical source. When the combustion gases in the room reach the predetermined concentration, trigger the alarm and turn off the general electrical source, thereby achieving an explosion-proof state.

The other type of logging unit house is non-explosion-proof. All facilities in the room are not explosion-proof. When this house is arranged, it should be more than 30 m from the wellhead (Fig. 10-2-12).

FIG. 10-2-9 Lightening apparatus in the generating house.

FIG. 10-2-10 Explosion-proof and non-explosion-proof electric devices in the geo-house.

FIG. 10-2-11 Positive pressure logging house.

FIG. 10-2-12 Non-explosion-proof logging house about 50 m from the wellhead.

Hidden danger: The bolts on the shell of the explosion-proof electric devices don't screw tightly

Hazard

The bolts on the shell of the explosion-proof electric devices don't screw on tightly. The explosion-proof property of the electric devices will be invalid (Figs. 10-2-13, 10-2-14).

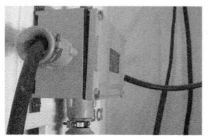
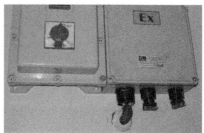

FIG. 10-2-13 The bolts on the separated explosion shell don't fit tightly.

FIG. 10-2-14 The bolts on the separated explosion shell loosened and don't fit tightly.

Remedy

Install and use the explosion-proof devices per product instructions. Tighten all bolts on the separated explosion shell.

Defect: Explosion-proof electric devices are used beyond the explosion-proof area

Hazard

Outside of the explosion-proof area is considered a nondangerous area. Although the explosion-proof electric devices used in the nondangerous area aren't harmful, they are much more expensive than non-explosion-proof electric devices. It is not necessary to use explosion-proof electric devices in the nondangerous area, and it is waste of resources (Figs. 10-2-15 to 10-2-18).

FIG. 10-2-15 The explosion-proof lights, connectors in the generating house.

FIG. 10-2-16 The explosion-proof air-conditioning in the cadre duty house.

FIG. 10-2-17 The explosion-proof lights in the roughneck duty house.

FIG. 10-2-18 The explosion-proof connector in the geo-house.

Remedy

Use of ordinary electric devices out of the explosion-proof area can meet jobs completely. Even if some situations involve explosion-proof electric devices, there is no way to realize that the system is explosion-proof; it simply is excessive. For example, there is flame in the boiler house, and the generator itself produces electric sparks, but computers, printers, and a majority of equipment in the logging and geo-houses is not explosion-proof. Even if these situations are equipped with explosion-proof switches, lights, connectors, and such, it will not make the whole system explosion-proof.

Hidden danger: The gas relief pipe of the diesel engine has no cooling and spark-eliminating devices

Hazard

After drilling into an oil and gas zone, natural gas may overflow from the well at any moment. If the gas relief pipe of the diesel engine has no cooling and spark-eliminating devices (Figs. 10-2-19, 10-2-20), the gas relief pipe will get hot, and sparks spouted from the pipe will light up the explosive mixture that is all around.

FIG. 10-2-19 There are no cooling and spark-eliminating devices (A).

FIG. 10-2-20 There are no cooling and spark-eliminating devices (B).

Remedy

After drilling into an oil and gas zone, there is a potential for an explosion around the derrick floor, including oil and gas cut, overflow, kick, and blowout; the cooling and the spark-eliminating devices of the gas relief pipe must be started. If not drilling into an oil and gas zone, or not discovering oil and gas cut, you can't start the cooling and spark-eliminating devices, but you must ensure that once they need to be started, they can be started at once. Figures 10-2-21 through 10-2-24 show the recommended coolers and spark-eliminating devices.

At present, water is used to spout the gas relief pipe or a water tank is mounted outside the gas relief pipe to cool the gas relief pipe of the diesel engine on the drilling site.

FIG. 10-2-21 Cooling and spark-eliminating devices (A).

FIG. 10-2-22 Cooling and spark-eliminating devices (B).

FIG. 10-2-23 Cooling and spark-eliminating devices (C).

FIG. 10-2-24 Cooling and spark-eliminating devices (D).

Defect: The diesel engine's cooling and spark-eliminating devices increase power loss

Hazard

The cooling and the spark-eliminating devices are used by some drilling crews, and the gas relief pipe is made very long, with many curves or bent portions in it (Figs. 10-2-25, 10-2-26). Although, in this way, it can function to cool and eliminate sparks of the gas relief pipe, it can produce greater back pressure in the pipe, more than 2 MPa (*SY/T 6586—2003*). Not only does this influence the gas releasing, but it also increases the power loss of the diesel engine.

FIG. 10-2-25   Longer gas relief pipe and more bends (A).

FIG. 10-2-26   Longer gas relief pipe and more bends (B).

Remedy

The spark-eliminating device can be also called a spark extinguisher or shield, fireproofing cap, and so on. It allows air current to pass the pipe, holds the flame and sparks to the spout. They can be divided into dry-type spark extinguishers and water-based spark extinguishers. Dry-type spark extinguishers are also divided into two types: spread-type and extinguishing-type. The spread-type spark extinguisher can remove the hot micro particles from the air current spouted by turbulence. The extinguishing-type spark extinguisher can separate and extinguish the hot micro particles, and then gas relief them into the atmosphere safely. On the drilling site, generally the dry-type spark extinguisher is used. Almost all drilling crews use spark extinguishers similar to that in Figure 10-2-24. This spark extinguisher only has one servicing state; that is, in spite of drilling in an oil and gas zone and eliminating sparks, the gas relief pipe of the diesel engine produces back pressure all the time while the engine is servicing or gas releasing. This device will lead to a greater power loss of the diesel engine and a higher consumption of diesel oil.

If only there were a spark extinguisher, it could not only provide fireproofing and explosion-proofing, but also the power loss of the diesel engine is not increased, and it decreases the consumption oil of diesel in the course of normal production. The spark extinguisher is open when the oil pool has been drilled in, and does not produce the back pressure when gas releasing. But when the spark extinguisher is used because of oil-gas cut or overflow, it is closed (production of back pressure), eliminating sparks. Figures 10-2-27 and 10-2-28 show an adjustable spark extinguisher. The spark extinguisher can be closed or opened by pulling a rope connected to the switch.

FIG. 10-2-27   Adjustable   spark   extinguisher (A).

FIG. 10-2-28   Adjustable   spark   extinguisher (B).

Hidden danger: The gas relief pipe of the diesel engine is toward the wellhead or circulating tank area

Hazard

The wellhead and circulating tank area are the places where natural gas gathers easily, and they are dangerous areas. There are flames and sparks in the air current gas released by the gas relief pipe of the diesel engine. If the pipe is toward the dangerous areas—that is, if it points toward the wellhead or mud tank area—it increases the explosive severity (Fig. 10-2-29, 10-2-30).

FIG. 10-2-29 The diesel gas relief pipe is toward the wellhead.

FIG. 10-2-30 The diesel gas relief pipe is near the wellhead.

Remedy

Although the distance between the gas relief pipe of the diesel and wellhead is less than 15 m (*SY/T 6426—2005*), the gas relief pipe of the diesel should be further, and should escape from the wellhead and mud tank area (the dangerous area) as quickly as possible. The diesel gas relief pipe can point to the left or the rear. In order to install the cooling and spark-eliminating devices, the pipe can't be upright and up toward the dangerous area (Figs. 10-2-31, 10-2-32).

FIG. 10-2-31 Gas relief pipe points to the left of the well site.

FIG. 10-2-32 Gas relief pipe points to the rear of the well site.

Hidden danger: The boiler house is located downwind

Hazard

The boiler house is a big fire source. When an overflow or blowout happens, the natural gas spilt over the wellhead drifts with wind to the boiler house, making the boiler explode. In Figures 10-2-33 and 10-2-34, the wind blows toward the well site from the left front, and the boiler house is located on the leeward.

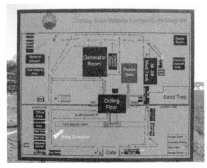
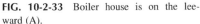

FIG. 10-2-33   Boiler house is on the lee-   **FIG. 10-2-34**   Boiler house is on the leeward (A).   ward (B).

Remedy

It is a mistake to install the boiler on the leeward of the seasonal wind; it must be installed on the windward (Figs. 10-2-35, 10-2-36). One point of view thinks that if the boiler is on the windward of the seasonal wind, the sparks coming from the chimney of the boiler drift to the wellhead and light the natural gas spouting out of the well. It is impossible that sparks from the chimney of the boiler drift over 50 m to the wellhead. Even if the sparks drift to the wellhead beyond 50 m, it can't have the igniting energy.

For the position of the boiler, there are different requirements for different standards. *SY/T 5225—2005* states that the "boiler house should be on the windward"; *SY 5466—2004* states that the "boiler house should be on the leeward of season wind"; and *SY5974—2007* states that the "boiler house should be on the leeward, and the distance from wellhead is greater than or equal to 50 m." But *Oil And Gas Engineering Design Fireproofing Specifications GB 50183—2004* regulate definitely that the fire source is not to be located on the leeward of seasonal wind. I think the requirements of *GB 50183—2004* about the position between fire and leakage sources are right.

When selecting the direction of the V-door of the derrick, in order to avoid the combustion gases overflow from the wellhead drifting to the generating house or the power equipment parked in the well site, causing an explosion, the seasonal wind should blow toward the well site from the left front. So the drilling burning coal boiler should be installed on the left front of the well site.

It is inadvisable that the spark extinguisher be installed on the boiler chimney. The boiler is put in the non-explosion-proof area 50 m or more from the wellhead. Even though the chimney spouts sparks, they can't ignite the natural gas overflowed from the wellhead. If the spark extinguisher is installed on the boiler chimney, it will influence the efficiency of burning coal.

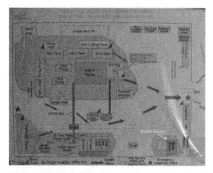 

FIG. 10-2-35 Boiler house is on the windward (A).

FIG. 10-2-36 Boiler house is on the windward (B).

Hidden danger: The distance between the oil tank and generating house is less than 20 m

Hazard

The oil tank area is flammable and an explosive man-trap. Once the oil tank leaks, the fire source of the generating house is likely to ignite the leaked diesel oil and cause a fire or explosion because of the distance between the generating house and oil tank (Fig. 10-2-37).

Remedy

The distance between the oil tank area and generating house should not be less than 20 m (Fig. 10-2-38).

FIG. 10-2-37 The distance between the oil tank and generating house is about 5 m.

FIG. 10-2-38 The distance between the oil tank and generating house is more than 20 m.

3.   SAFETY MEASURES OF PREVENTING HYDROGEN SULFIDE

Hidden danger: There is too great a distance between the hydrogen sulfide monitor and the monitoring level

Hazard

Hydrogen sulfide is a little heavier than air; its relative density is 1.19. Hydrogen sulfide gets together in the low-lying region easily. If the hydrogen sulfide monitor (detector) is too far from the monitoring level, it can't monitor the hydrogen sulfide concentration correctly (Figs. 10-3-1, 10-3-2).

FIG. 10-3-1   Monitor is about 1.6 m from the derrick floor.

FIG. 10-3-2   Monitor is about 1.5 m from the derrick floor.

Remedy

When the relative density of gas is greater than 0.97, the installation of the monitor should be 0.3 to 0.6 m from the monitoring level. The relative density of hydrogen sulfide is 1.19, so in order to predict the concentration of hydrogen sulfide correctly, the installation of the monitor should be 0.3 to 0.6 m from the monitoring level (Figs. 10-3-3, 10-3-4).

FIG. 10-3-3   The monitor is about 0.4 m from the level of the tank.

FIG. 10-3-4   The monitor is about 0.4 m from the derrick floor.

Hidden danger: The non-explosion-proof alarming controller of hydrogen sulfide is put in the explosion-proof area

Hazard

Unless there are special requirements, the alarming controller of hydrogen sulfide (also called a quadratic meter) commonly is not explosion-proof. If you put the non-explosion-proof alarming controller in the explosion-proof area, it will make the whole system lose its explosion-proof function (Figs. 10-3-5, 10-3-6).

FIG. 10-3-5   Controller is hung in the duty house.

FIG. 10-3-6   Controller is in the dog house.

Remedy

The non-explosion-proof alarming controller should be in the non-explosion-proof area at least 30 m from the wellhead, and in the duty house where a person is on duty. It can be in the cadre duty house. If the controller is explosion-proof, it should be near the driller's console or in the duty house. It is convenient for people on the floor and people on duty to know the value of hydrogen sulfide in time (Figs. 10-3-7, 10-3-8).

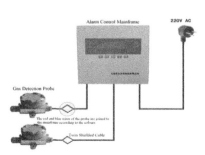

FIG. 10-3-7   Non-explosion-proof alarming controller is hung in the cadre duty house.

FIG. 10-3-8   Explosion-proof controller in the duty house.

Hidden danger: The hydrogen sulfide sign tag doesn't provide the status of the well site

Hazard

Figure 10-3-9 shows a caution tag with three colors, green, yellow, and red. The current status of the well site can't be estimated from the tag. Whether there is hydrogen sulfide, whether there is a danger to life or health, and the circumstances, is not clear in Figure 10-3-10.

FIG. 10-3-9   Caution tag unknown (A).

FIG. 10-3-10   Caution tag unknown (B).

Remedy

There should be a distinct, clear caution sign at the well site where it is likely to encounter hydrogen sulfide. If the concentration of hydrogen sulfide is less than 15 mg/m^3 (10 ppm), the green tag should be hung; if the concentration of hydrogen sulfide is 15 mg/m^3 (10 ppm) to 30 mg/m^3 (20 ppm), the yellow warning should be hung; if the concentration of hydrogen sulfide is greater or may be greater than 30 mg/m^3 (20 ppm), the red warning should be hung. Don't allow three colors on the warning board at once. The warning boards in Figures 10-3-11 and 10-3-12 are used for reference.

FIG. 10-3-11   Distinct and clear warning sign (A).

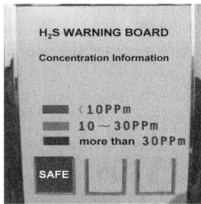

FIG. 10-3-12   Distinct and clear warning sign (B).

Defect: A warning sign is hung in a well site without hydrogen sulfide

Hazard

SY/T 5087—2005 regulates that "There is a distinct and clear warning sign in the operation well site where hydrogen sulfide maybe encountered." A warning board is not necessary if it is impossible to encounter hydrogen sulfide in the operation well site. The meaning of a green tag is to point out that "the well is in control, but there is a potential danger to life and health." For a well without hydrogen sulfide, a green tag is misleading people may think the well contains hydrogen sulfide, but it just doesn't exceed 15 mg/m^3 (Figs. 10-3-13, 10-3-14).

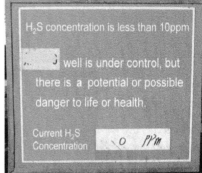

FIG. 10-3-13 Warning board is hung in a well site with no hydrogen sulfide (A).

FIG. 10-3-14 Warning board hung in a well site with no hydrogen sulfide (B).

Remedy

It is inadvisable for a warning board for hydrogen sulfide to be hung in a well site with no hydrogen sulfide.

Defect: The facilities room contains all positive pressure respirators

Hazard

Positive pressure respirators are put in the facilities room more than 30 m from the wellhead. When an overflow or kick arises, the concentration of hydrogen sulfide exceeds 30 mg/m^3 (20 ppm), workers rush into the facilities room, put on respirators, and then return to their posts, completing the shut-in process to control the leakage of hydrogen sulfide. The shut-in time will be delayed by 5 to 10 minutes at least, making it difficult to control the leakage in time. It is likely that the wellhead should have been controlled, but it shouldn't have been controlled.

Remedy

Put the positive respirators in the dog house or duty house, where drilling personnel can fetch and use them rapidly (Fig. 10-3-15, 10-3-16). When hydrogen sulfide concentration exceeds 30 mg/m^3 (20 ppm), put on the respirator rapidly and complete the shut-in process to control leakage and avoid harm.

FIG. 10-3-15   Air respirators in the duty house.   **FIG. 10-3-16**   Air respirators in the dog house.

Hidden danger: There is no breath air compressor

Hazard

Generally, the use time of a bottle of positive pressure respirator is about 30 minutes. If the circumstances on site are complex, and many people need to operate continuously, if there is no breath air compressor, air cannot be charged into the air bottle, and it will influence continuity of operation.

Remedy

When a drilling operation is in a sulfur-bearing region, the breath air compressor should be available (Figs. 10-3-17, 10-3-18). The compressor should be outside the explosion-proof and clean air area.

FIG. 10-3-17   Breath air compressor (A).   **FIG. 10-3-18**   Breath air compressor (B).

Chapter 11

Other Well Control Hidden Dangers and Remedies

There are several common hidden dangers for well control devices:

- The bolt is shorter on one side of the flange connection.
- There are different standards for flange bolts.
- The blowout preventer pipe, relief pipe, drilling fluid recycling pipe, and fluid feed pipe for the gas-fluid separator are field welded.
- There is no cutoff valve for the manifold pressure gauge.
- The fingers of the pressure gauge fall off or cannot return to zero.
- The well control devices were not painted in a warning color (e.g., red).

The frequent hidden dangers for shutting-in warning signs are:

- The names of these signs have defects.
- There is no recommended value for the shut-in casing pressure corresponding with different fluid densities.
- The warning signs are located away from the choke valve, which is to be operated manually.

Common defects for the fluid level alarm are:

- The mechanical level ruler is calibrated in length unit such as centimeter.
- The maximum interval value for the scale plate is not an integral multiple of cubic meters (m^3).
- The drilling fluids circulating tank didn't install a scale plate.
- The scale plates are immovable.
- False alarms occur frequently.
- A straight ruler is used to measure the level of drilling fluid.
- The scale plates are installed inside the pots.

Common hidden dangers for the alarm loudspeaker at the drilling floor are:

- The automatic reset button is used.
- The loudspeaker isn't loud enough.

Common defects for blowout prevention practice are:

- The slab gate valve, after the cutoff valve, stays closed during the practice.
- People use hand signals for communication.

Other hidden dangers include:

- Data from the duty recode is inconvenient for calculating the level variation of the drilling fluid.
- Pressure gauges are irrelevant.

1. WELL CONTROL DEVICE

Hidden danger: The bolt is shorter on one side of the flange connection

Hazard

There are two major hidden dangers when a bolt is shorter on one side of the flange connection:

1. The fasten force from the bolt is not sufficient, which causes the bolt to stay in the spring zone of the materials. In this situation, it is impossible to keep a good seal within the flange connection when concertina movement occurs.
2. If one side of a bolt is too short for its nut, the worker may keep twisting the bolt. This situation can overstress the bolt, leading to a transfer from spring zone to plasticity zone. This transfer will buckle the bolt permanently, reducing or losing seal capability.

There are three reasons that cause a bolt to be shorter on one side:

1. The length of the bolts is shorter than the recommended length; therefore the bolt cannot meet the nut for two units on both sides.
2. The worker didn't twist the bolt well, making one side too long and the other too short.
3. The back-up torque is too weak.

As shown in Figures 11-1-1 to 11-1-4, the shorter bolt on one side of the flange connection can be found in the blowout preventer (BOP), BOP manifold, well killing throttle manifold, relief pipe, drilling fluid recovery pipe, and liquid-free pipe of the liquid-gas separator.

FIG. 11-1-1 Bolt shorter in one side of the blowout preventer flange.

FIG. 11-1-2 Bolt shorter in one side of the spool flange.

FIG. 11-1-3 Bolt shorter in one side of the blowout preventer manifold flange.

FIG. 11-1-4 Bolt shorter in one side of the throttle pipe.

Remedy

The length of the bolts should be designed to accommodate the thickness of the fastening components and the nuts. More specifically, when the nut is fixed, the bolt should be two or three units over the nut, which is the recommended value. When using the recommended bolts, it is easy to observe if the bolt and nut are fastened properly (in other words, not under- or overtwisted), judging by the length that is left over. Observing both sides of a fastened bolt can ensure that it is fastened effectively (Figs. 11-1-5 to 11-1-8).

FIG. 11-1-5 Spool flange bolts are over the nut by 2 or 3 threads.

FIG. 11-1-6 Fastened blowout preventer bolts are over the nut by 2 or 3 threads.

FIG. 11-1-7 Fastened killing well manifold bolts are over the nut by 2 or 3 threads.

FIG. 11-1-8 Fastened blowout preventer manifold bolts are over the nut by 2 or 3 threads.

Hidden danger: There are different standards for flange connection bolts

Hazard

If different flange bolts and nuts are used, the moment of link stress may not be unified. This situation can cause degradation or even loss of function for the flange connection (Figs. 11-1-9, 11-1-10).

FIG. 11-1-9   The thickness of the flange nuts is different.

FIG. 11-1-10   The length of the flange bolts is different.

Remedy

The same standard for flange connections should be adopted. The length and thickness of the flange bolts and nuts should be unified. When fastened, the flange bolt should be over the nut by two or three threads.

Hidden danger: Field welding the blowout preventer choke/kill line and flow line pipe, drilling fluid recovery pipe, and fluid feed pipe for the gas-fluid separator

Hazard

It is common to field weld the blowout preventer pipe, relief pipe, drilling fluid recovery pipe, and fluid feed pipe, as shown in Figures 11-1-11 to 11-1-14 (especially for the latter two pipes). Field welding is difficult because well control pipes are highly stressed. At present, there is not a feasible approach for field welding on a specified pipe when considering the materials and the use requirement. Thus the welding parameters, such as current, voltage, the power of the welding wire, and whether pre- or post-heat is needed to avoid welding stress, are not determined. On the other hand, real time monitoring during welding cannot be implemented. Thus the final quality assurance procedure cannot be completed, which includes checking the shape of the welding structure, inspecting the welding gap, a lossless check of the welding head, sealing check of the welding head, and stress endurance of entire structure. Consequently, quality assurance cannot be guaranteed.

In many oil fields, stabbing and draining disasters occasionally happened when welding a pipe under high pressure or even much lower than the rated pressure.

Remedy

Field welding for well control pipes is forbidden.

To facilitate the connection of the drilling fluid recovery pipe and hence avoid additional welding, high-stress, fire-resistant soft pipe should be adopted as the recycling pipe.

FIG. 11-1-11   The relief pipe in field welding.

FIG. 11-1-12   Drilling fluid recycling pipe in field welding.

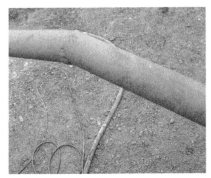

FIG. 11-1-13   The drilling fluid recycling pipe in field welding.

FIG. 11-1-14   The fluid feed pipe for the separator in field welding.

Hidden danger: There is no cutoff valve for the manifold pressure gauge

Hazard

If there is no cutoff valve beneath the manifold, the pressure gauge cannot be replaced when pipes endure high stress (Fig. 11-1-15).

Remedy

To be able to replace the pressure gauge under stress, a cutoff valve should be installed beneath the pressure gauge (Fig. 11-1-16).

FIG. 11-1-15 No cutoff valve is installed beneath the pressure gauge.

FIG. 11-1-16 A cutoff valve is installed beneath the pressure gauge.

Hidden danger: The fingers of the pressure gauge fall off or cannot return to zero

Hazard

When fingers of the pressure gauge fall off or cannot return to zero, the pressure gauge will lose its functions, as illustrated in Figures 11-1-17 and 11-1-18.

FIG. 11-1-17 Fingers of the pressure gauge fall off.

FIG. 11-1-18 Fingers of the pressure gauge cannot return to zero.

Remedy

The installed pressure gauge should be checked annually. Additionally, during the drilling period, in order to guarantee that the pressure gauge works properly, they should be checked frequently, at least once daily. Things to check daily include: (1) whether the fingers of pressure gauges are in the correct location; (2) whether damping oil is adequate; (3) whether the glass that covers the surface is undamaged.

Defect: The well control devices were not painted with a warning color (e.g., red)

Hazard

Though the devices have different functionalities, they should be all painted in safety colors. In an emergency situation, workers can react properly based on their knowledge of safety colors. These reactions include recognizing the dangerous parts, fast response, improving self-control ability, and reducing the risk. The industry standard for the warning color for well control devices is red. Currently, devices like BOPs, throttle manifolds, blow-off manifolds, and gas-fluid separators are painted in other colors, illustrated in Figures 11-1-19 and 11-1-20.

FIG. 11-1-19 Well killing manifolds are painted blue and yellow.

FIG. 11-1-20 Liquid-gas separator is painted blue.

Remedy

The red color represents forbidden, stop, fire control, and dangerous. Every field or device that involves forbidden, stop, or dangerous should be painted red. The well control device with its manifold endures high stress, so it is a dangerous device, and should be painted in red. Moreover, the industry standard suggests that the well control device with its manifold be painted red, illustrated in Figures 11-1-21 to 11-1-24.

FIG. 11-1-21 Choke manifold is painted red.

FIG. 11-1-22 Killing line is painted red.

FIG. 11-1-23 Blowout preventer, spool, and the blowout preventer lines are painted red.

FIG. 11-1-24 Liquid-gas separator is painted red.

2. SHUT-IN WARNING SIGNS

Defect: The names of these warning signs have pitfalls

Hazard

Names like "Well Control Pressure Signboard," "Well Killing Signboard," "Well Control Warning Signboard," and "Well Control Working Signboard" are ambiguous, and can easily cause misunderstanding and misinterpretation. Also, those names are not strongly related to their role, which is to display the maximum allowed shut-in casing pressure, illustrated in Figures 11-2-1 to 11-2-4.

Remedy

The maximum allowed shut-in casing pressure warning signboard should be hung near the choke manifold. This signboard is used to remind the worker that the shut-in casing pressure should not be over the maximum allowed shut-in casing pressure to prevent breaking the ground or the casing pipes. The preferred names in the working field can be "Maximum Well Closing Stress," "Well Closing Stress," "Safety Well Closing Stress," or "Well Closing Signboard," illustrated in Figures 11-2-5 and 11-2-6.

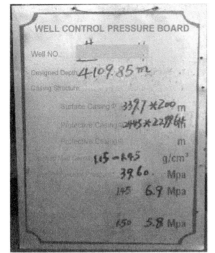

FIG. 11-2-1 Naming in the signboard is ambiguous (A).

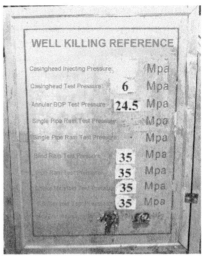

FIG. 11-2-2 Naming in the signboard is ambiguous (B).

FIG. 11-2-3 Naming in the signboard is ambiguous (C).

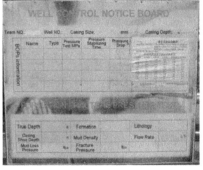

FIG. 11-2-4 Naming in the signboard is ambiguous (D).

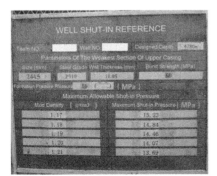

FIG. 11-2-5 The acceptable name for a warning signboard (A).

FIG.11-2-6 The acceptable name for a warning signboard (B).

Defect: No recommended value of allowable shut-in casings corresponds with different fluid densities

Hazard

The warning signboards shown in Figures 11-2-7 and 11-2-8 only displayed the allowed values of shut-in casing pressure under the maximum fluid density condition. However no specified maximum allowable shut-in casing pressure that is based on the different real drilling fluid densities is displayed. The real fluid density could be more or less than the theoretical maximum drilling fluid density. In Figure 11-2-8 the parameters such as "real casing program," "pressure coefficient of oil reservoir," and "the test pressure for the well control device" are irrelevant to the maximum allowed shut-in casing pressure, which should be deleted from the signboard, to prevent them from confusing the workers.

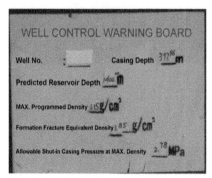

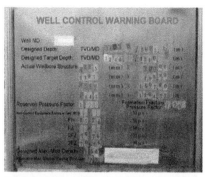

FIG. 11-2-7 Insufficient information is displayed on the warning signboard (A).

FIG. 11-2-8 Insufficient information is displayed on the warning signboard (B).

In Figures 11-2-9 and 11-2-10, only one maximum allowed shut-in casing pressure has been displayed, where the maximum allowed shut-in casing pressure corresponding to real different drilling fluid densities also need to be displayed.

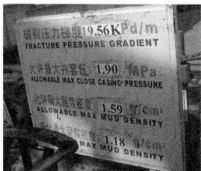

FIG. 11-2-9 Insufficient information is displayed on the warning signboard (C).

FIG. 11-2-10 Insufficient information is displayed on the warning signboard (D).

Remedy

In order to obtain the data precisely, the maximum allowed shut-in casing pressure, which is displayed on the signboard, should be updated to refer to the drilling fluid density (interval value) in the next drilling action (Figs. 11-2-11 to 11-2-14). The minimum interval for the changes of the drilling fluid can be around 0.05 g/cm^3. When the change occurs within the interval value, the current stress value on the warning signboard can be kept without modification.

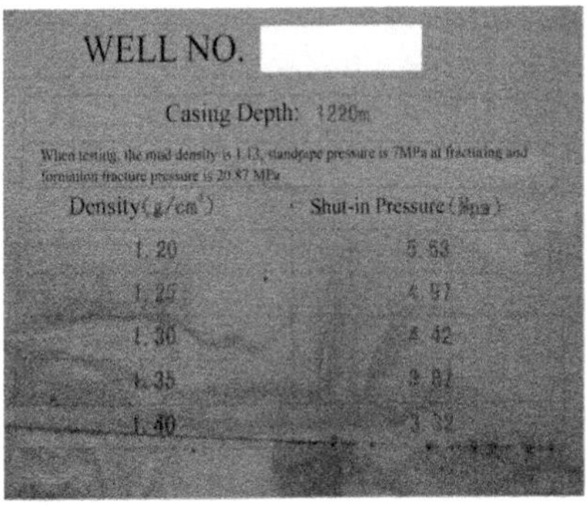

FIG. 11-2-11   Recommended warning signboard (A).

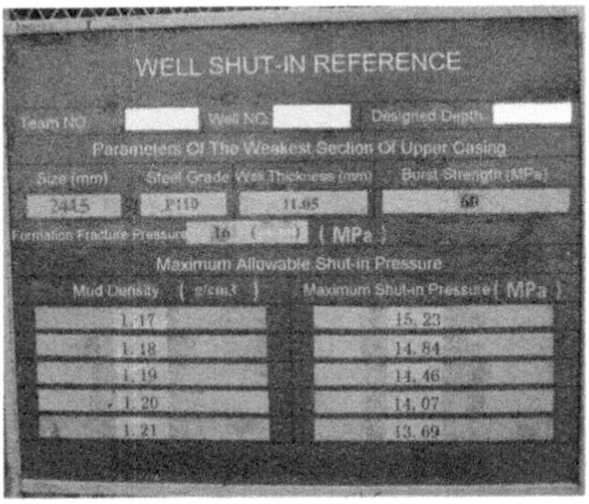

FIG. 11-2-12   Recommended warning signboard (B).

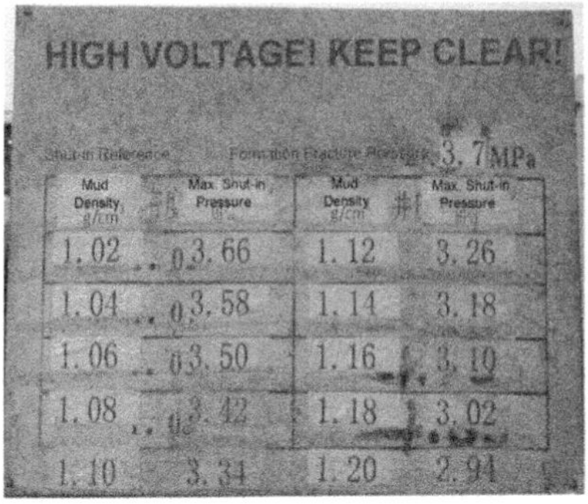

FIG. 11-2-13   Recommended warning signboard (C).

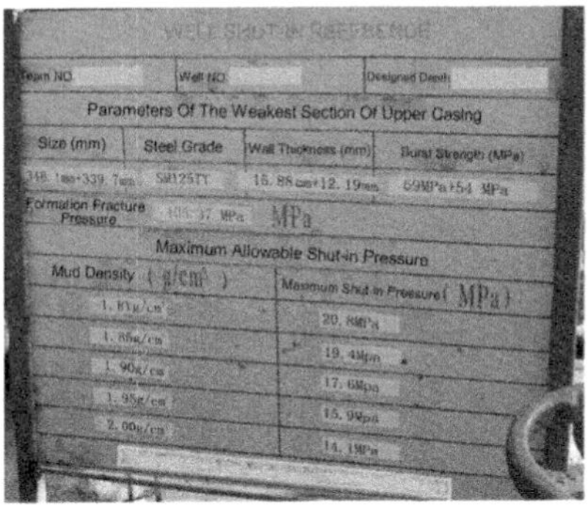

FIG. 11-2-14   Recommended warning signboard (D).

Defect: The warning signboards are located away from the manual choke valve

Hazard

If the warning signboard is located away from the throttle it is inconvenient for the worker to read the maximum allowable shut-in casing pressure when operating the choke valve (Figs. 11-2-15, 11-2-16).

FIG. 11-2-15 The warning signboard is away from the operator.

FIG. 11-2-16 The warning signboard is behind the operator.

Remedy

The warning signboard is best located in front of or near the manual choke valve, where the operator can easily read the closing value from the warning signboard when he looks up (Figs. 11-2-17, 11-2-18).

FIG. 11-2-17 The warning signboard is near the manual choke valve.

FIG. 11-2-18 The warning signboard is in front of the operator.

3. LIQUID LEVEL ALARM DEVICE

Defect: The mechanical liquid level ruler is measured in centimeters

Hazard

The operator on duty will read the volume of drilling fluid in the tank one or more times every 15 minutes, and can get the varied volume of drilling fluid only by first converting from centimeter to stere. This increases the extra work load of the operator on duty because of the need to perform the conversion, and it is easy to make mistakes due to tiredness. The mechanical liquid level scale is measured by centimeters or other length units.

This ruler is easy to fabricate by drawing a scale on it in centimeters, but it is not easy to use. So after reading the fluid level of the tank, we must check the conversion list for the drilling fluid tank's volume to get the volume of drilling fluid.

The crew members repeat such conversion work will easy to be fatigue and make mistakes, which commonly happens in many drilling crews (Figs. 11-3-1 to 11-3-4).

FIG. 11-3-1   Scale ruler graduation is a long measure (A).

FIG. 11-3-2   Scale ruler graduation is a long measure (B).

			MUD TANK CUBIC METER CONVERSION						
			APPLICABLE TO 2# 3# 4# 5# (M³)						
0	10	20	30	40	50	60	70	80	90
0	1.80	3.60	5.40	7.20	9.00	10.80	12.60	14.40	16.20
0.18	1.98	3.78	5.58	7.38	9.18	10.98	12.78	14.58	16.38
0.36	2.16	3.96	5.76	7.56	9.36	11.16	12.96	14.76	16.56
0.54	2.34	4.14	5.94	7.74	9.54	11.34	13.14	14.94	16.74
0.72	2.52	4.32	6.12	7.92	9.72	11.52	13.32	15.12	16.92
0.9	2.70	4.5	6.3	8.1	9.9	11.7	13.5	15.30	17.10
1.08	2.88	4.68	6.48	8.28	10.08	11.88	13.68	15.48	17.28
1.26	3.06	4.96	6.66	8.46	10.26	12.06	13.86	15.66	17.46
1.44	3.24	5.04	6.84	8.64	10.44	12.24	14.04	15.84	17.64
1.62	3.42	5.22	7.02	8.82	10.62	12.42	14.22	16.02	17.82

FIG. 11-3-3   Drilling fluid conversion table (A).

Volume Change Sheet Of Active Mud Tank

When 1# and 2# tanks being used for circulation, the total volume is 176.41m³, each little scale of the level alarm is 6.5mm, and one little scale for volume is 0.49m³.

When 1# ,2 # and 3# tanks being used for circulation, the total volume is 254.44m³, each little scale of the level alarm is 6.5mm, and one little scale for volume is 0.72m³.

Each little scale of the Metering Tank is 27mm, and one little scale for volume is 0.18m³.

Note: 1# shale shaker tank, 2# suction tank, 3# mixing tank

FIG. 11-3-4   Drilling fluid conversion table (B).

Remedy

The mechanical liquid level ruler should be scaled by volume unit cubic meter, so that the operator on duty can get the volume of drilling fluid in the trip tank from the mechanical liquid level ruler directly. This makes it easier to get the drilling fluid volume in the trip tank, and minimizes mistakes in measurement and conversion (Figs. 11-3-5, 11-3-6).

This kind of ruler can be scaled in stere units according to the conversion maximum for one cubic meter after measuring the tank's acreage. And it is easy to use except for the complicated fabrication process. Meanwhile, the manager should simplify some complicated works and reduce the operator's workload as much as he can. Execution can also be advanced by working simply, while errors and mistakes will be reduced.

FIG. 11-3-5 The scale was measured by volume unit (A).

FIG. 11-3-6 The scale was measured by volume unit (B).

Hidden danger: The scale on the mechanical liquid level ruler was not rounded to times of stere

Hazard

Compared with the mechanical liquid level ruler in length unit, the one in the volume unit can be read intuitively, conveniently, and clearly, and there is no need to convert. Some drilling crews marked the drilling fluid volume corresponding to 10 cm on the ruler for convenience when 10 cm was used as one grid. But it is not convenient to read data because the volume value of one grid was not rounded to integers, although the scale can also denote the volume value in this scaling way. In Figure 11-3-7, the big grid means 1.3 m³ while it means 1.4 m³ in Figure 11-3-8, and the small grid means 0.14 m³. And it is not convenient for the operator to observe and read values compared with using round figures like 1 m³.

FIG. 11-3-7 One big interval of scale means 1.3 m³.

FIG. 11-3-8 One big interval of scale means 1.4 m³.

Countermeasure and suggestions

The mechanical liquid level ruler should be easy and clear to read values. For example, while one big scale means 1 cubic meter or 2 cubic meters, and the small one means 0.5 cubic meters or 1 cubic meter, it is easy to get the value and can also fulfill the precision requirements for well control. Keep one accurate figure after decimal place can meet the demands for easy to detect a kick in time. (Figs. 11-3-9, 11-3-10). The scale of the mechanical liquid level ruler will fulfill the requirement to detect overflow when showing one decimal place.

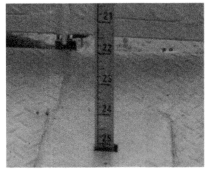

FIG. 11-3-9   One big interval of scale means 1 m³ (A).

FIG. 11-3-10   One big interval of scale means 1 m³ (B).

Hidden danger: The drilling fluid tank in circulation does not have a fluid level ruler

Hazard

The drilling fluid tank in circulation does not have a fluid level ruler.

The tank in circulation cannot monitor the volume or variety of drilling fluid exactly, which results in failing to detect an overflow in time.

Remedy

In order to inspect the sum volume variety of drilling fluid exactly, every tank should be equipped with a fluid level ruler that participates in circulation (Fig. 11-3-11).

FIG. 11-3-11   Every tank should be equipped with a fluid level ruler (the red pole on the tank).

Hidden danger: The mechanical liquid level ruler is not flexible

Hazard

The operator on duty cannot observe an overflow because if the mechanical liquid level ruler is not flexible, it will fail to float with the fluid level, and will fail to indicate the volume variety in time.

Remedy

The mechanical liquid level ruler should keep its flexibility during use to ensure that the scale will float with the fluid level and indicate the volume variety exactly and in time.

Defect: The fluid level alarm device causes a false alarm

Hazard

Most of the drilling crews are using the alarm device installed on the mechanical liquid level ruler's bracket, which has a $\pm 1\ m^3$ limitation. When the volume variety exceeds $1\ m^3$, the alarm device will sound a long buzzer the same as the speaker that cautions all the people on the drilling site (Figs. 11-3-12, 11-3-13).

FIG. 11-3-12 Liquid quantity variation alarming limitation being $1\ m^3$ (A).

FIG. 11-3-13 Liquid quantity variation alarming limitation being $1\ m^3$ (B).

There will always be some drilling fluid remaining in the mud flume or trough and in the circulation trough that is moving into the tank constantly, increasing the volume variety over $1\ m^3$ after switching off the pump every time. After switching on the pump, the drilling fluid will not get out from the well immediately because of the time lag, and it will also take some time for the drilling fluid flowing into the drilling fluid tank from the circulation trough, which reduces the drilling fluid volume clearly to 3 to 5 m^3 or 9 m^3 sometimes before restoring to normal. And the fluid volume variety will exceed $1\ m^3$ when the pump is switched on or off, while the exact value is related to the size of the well, and the depth, quantity, and swept volume of the volume and performance of the drilling fluid.

The alarm device will sound a long buzzer during the early time of switching on or off, which can be heard by all operators on the drilling site. The switch on or off of the pump is a normal operation, so the alarm will always be triggered, which makes operators become impervious to it. And possibly no one will attach importance to the alarm that indicates that the well should be closed in time when the overflow and well kick really happen.

Remedy

It's improper to set the alarm value greater than 1 cubic meter too. So this kind of alarm device always sends out wrong alarm signals. If the alarm device is removed, just leaving the liquid level ruler, can only rely on operators reading the value every 15 minutes to inspect and alarm. This is only an expedient measure.

The liquid ruler should be used to read the values and also have the function to alarm. And it is suggested that the impulse bourdon electric bugle should be used together with an impulse warning light that can be installed on top of the duty room or beside the operation platform separately. The electric bugle is different from the sound of the well control warning bugle, which should not be high. The sound of the electric bugle and flicker of the warning light will warn the operator and driller that the drill fluid changes when the volume varied exceeds $1 \, m^3$. Later, the operator and driller will confirm the reason for inducing drilling fluid variety. If it is while the pump is switching on or off, the alarm will be unchanged while the driller will use the drilling control warning bugle on the platform to sound a long alarm to stop a process's implementation when a likely overflow is happening (Figs. 11-3-14, 11-3-15).

FIG. 11-3-14   Mechanical liquid level rule without the alarm function.

FIG. 11-3-15   Red warning light on top of the duty room.

Defect: Use the ruler to measure the fluid level

Hazard

It is inconvenient and inaccurate to measure the fluid level by ruler. The low visibility at night and rising steam inside the tank during the winter will affect the surveyor's sight (Fig. 11-3-16). Usually, the ruler is scaled by length unit. So the fluid level should be converted to volume to get the drilling fluid volume varied, which is a lot of work, is hard to calculate, and always causes mistakes after measuring the length between the fluid level and the surface of the tank. In Figure 11-3-17, the floodlight was installed to improve the visibility. (The floodlight was installed at a height of less than 10 cm on top of the tank surface, which can be easily damaged and can induce an electrical shock accident. The floodlight switch was forbidden to be installed less than 1.8 m under the tank surface.)

Remedy

Use a mechanical liquid level ruler and an electric fluid level to inspect the device at the same time.

FIG. 11-3-16 Using a ruler to measure the fluid level.

FIG. 11-3-17 Installed floodlight to make it easy to measure.

Hidden danger: The fluid level ruler is installed on the wall of the tank

Hazard

Some drilling fluid tanks were installed with a level ruler on their walls (Figs. 11-3-18, 11-3-19). The warm drilling fluid in the tank will emanate steam to affect the visibility, and the drilling fluid will also adhere to the scale, both of which will make it hard to read the scale.

Countermeasure and suggestions

Installing a mechanical liquid level ruler or electric level detect device on the wall provides agility, credibility, and accuracy.

FIG. 11-3-18   Level ruler is installed on the inside wall of the tank (A).

FIG. 11-3-19   Level ruler is installed on the inside wall of the tank (B).

4. WARNING BUGLE ON DERRICK FLOOR

Hidden danger: Using an automatic reset pneumatic bugle button

Hazard

With an automatic reset pneumatic bugle, when you press the button, it outputs an electric signal, which supplies gas to the bugle, and the bugle sounds. After the button loosens, the signal vanishes and the pneumatic bugle stops sounding. The driller cannot operate the brake crank to lift the Kelly or offload the drilling string and engage the inside blowout preventer quickly or perform other actions while one of his hands is pressing the button to sound for blowout (no less than 15 s while some standards call for no less than 30 s). The driller can only carry out other operations after stopping the warning signal, which will delay 15 to 30 s to shut-in (Figs. 11-4-1, 11-4-2).

FIG. 11-4-1   Automatic reset warning bugle button (A).

FIG. 11-4-2   Automatic reset warning bugle button (B).

Remedy

The remaining button should be used as the blowout warning button, which will allow the driller to operate the drawworks and brake crank to shut-in immediately while sending out the long alarm (Figs. 11-4-3, 11-4-4). After pressing it the first time, the button will continue to output the electrical signal, supplying the bugle with gas, so the bugle will continue to sound when the button is released. The bugle will stop after pressing the button a second time.

FIG. 11-4-3   The remaining button is switched off.

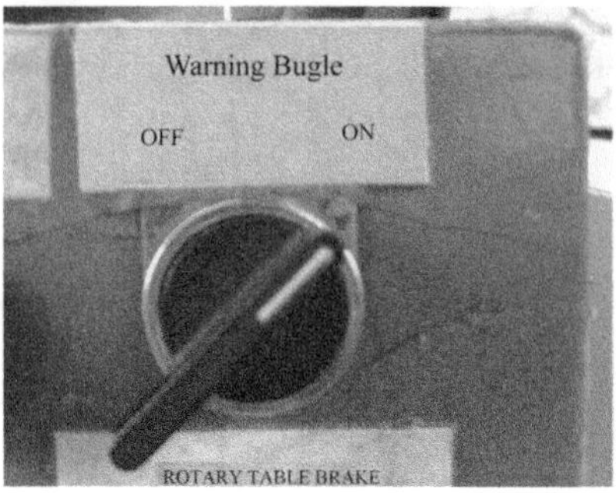

FIG. 11-4-4   The remaining button is switched on.

The driller will start the bugle, which will send out an alarm that is no less than 15 s (some standards call for no less than 30 s) first to signal all operators to prepare to close the well. Operations such as lift Kelly or offload drilling tool that engage the inside blowout preventer will not be affected when the driller emits the alarm.

Hidden danger: The bugle is not loud enough

Hazard

The bugle is not loud enough, because the bugle is faulty, so some of the drilling crew cannot hear the alarm. As a result, the well cannot be closed immediately because operators at some positions do not respond.

Remedy

Eliminate the faulty bugle to ensure all the operators at any position can hear the long alarm for blowout.

5.   COMMON DEFECTS AND REMEDIES OF BOP DRILLS

Hidden danger: During the BOP drill, the flat valve stays closed after turning off the throttling valve

Hazard

Some of the 70 MPa or 105 MPa throttle/choke manifolds (Figs. 11-5-1, 11-5-2) and the flat valve (J5, J7, J8, J9, and J10) behind the throttle valve have a lower loading capacity than the valve in front of the throttle valve. For this kind of throttle manifold, try to turn off the throttle valve, and then close the flat valve behind it. In this way, the J5 and J7 valves may take more pressure than the rated pressure, which is 35 MPa or 70 MPa.

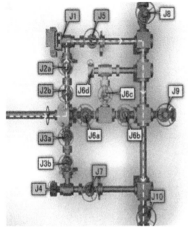

FIG. 11-5-1 Throttle manifold with 105 MPa (A).

FIG. 11-5-2 Throttle manifold with 105 MPa (B).

Of course, to the throttle manifold with the same pressure grade upstream and downstream, which is the user's specified requirement, turning off the aforementioned valve will not lead to overpressure operation.

Remedy

During BOP drills, after trying to turn off the throttle valve, you should close the gate valve (J2a or J3b) in front of the throttle valve, and then accomplish the whole shut-in program, even if overflow or blowout occurs.

Hidden danger: Using hand gestures to transfer signals during the BOP drill

Hazard

Some drilling crews use gestures to transfer signals instead of a loudspeaker during a BOP drill. There are many problems in using gestures:

1. During severe weather, such as dense fog, torrential rain, heavy snow, and sandstorms, people's sight is blurred; as a result, the gesture signal will be unclear.

2. When a blowout occurs, the borehole fluid that was blown from the wellhead will make it hard for the driller to see the gesture from the other side of the wellhead (Fig. 11-5-3).

3. When the driller receives the signal from the assistant driller and weevil, for which the equipment stands by and the shut-in condition is ready, he will issue the piccolo twice, which is the signal to close the gate, instead of issuing the closed gate signal at once. This delays the shut-in time. Some of the drilling crews have to wait for the gesture for 1 to 2 minutes, closing the gate after a long time (Fig. 11-5-4).

FIG. 11-5-3 Driller's sight is not clear after a blowout.   **FIG. 11-5-4** Using gestures to transfer signals before closing gate drills.

4. When the blowout occurs, it is hard for the crew to keep their composure, and it's impossible to transfer a sequence of gestures smoothly at this time.

Nowadays, gesture has been indispensable during BOP exercising. Gesture looks good but is not practical. How many people will gesticulate properly when a blowout happens?

Remedy

When the blowout occurs, only two workers are needed to close the well in tens of seconds if there are a driller's console and choke console. The most important thing is shutting down the well.

Simple and practical signals for shutting down the well are:

1. When the blowout occurs, the driller should issue a long horn, then each post should check the equipment to see if it's in the standby state or not. After that, the assistant driller should open the hydraulic valve.

2. When the drill floor has the condition of shutting-in, the driller issues the shutting-in signal immediately, which is two piccolo sounds. This piccolo sound means that it is time to shut down the BOP.

3. The assistant driller shuts down the BOP upon hearing two piccolo sounds, and then the roughneck tries to shut down the throttle valve, checks the pressure of the riser and casing, and closes the front flat valve of the throttle valve.

If there is a driller's console, the driller can open the hydraulic valve on the drill floor directly to shut down the BOP.

Notice: It is not necessary for the driller to wait for the assistant driller and weevil's gesture. If the drill floor is shutting down the well, the driller then issues two piccolo sounds immediately. The program of shutting down the well should begin with the driller.

6.   OTHER HIDDEN DANGERS AND REMEDIES

Hidden danger: The records on duty are not easy to calculate the variation of drilling fluid or there are some irrelevant parameters

Hazard

The operators on duty read the data of each liquid level gauge every 15 minutes, and then fill these data on records on duty. Then they calculate the amount of drilling fluid, contrast and analyze the variation of drilling fluid, and finally make the judgment of overflow or leakage.

In Figure 11-6-1, there is no trip column number, trip generation volume, or drilling fluid recruitment of normal drilling; as a result, it is hard to analyze the reason for the drilling fluid's variation.

MUD LEVEL MONITORING RECORD FORM

Mud Properties On Duty Shift Weight /./8   Viscosity 47   YY-MM-DD

Time	Working Condition	Well Depth	Mud VOL. In Each Tank (m³)				VOL. Change (m³)	Reasons Analysis	Ditch Indication	Observer
			1#	2#	3#	4#				
12:15	Drilling	3031.65	27.4	23.8	21.7	21.6	-0.5	Consumed	None	
12:30	Drilling	3032.18	27	23.7	21.2	21.6	-0.5	Consumed	None	
12:45	Circulation	3032.18	27	23.7	21.6	21.6	-0.1	Consumed	None	
13:00	POOH	3032.18	27	23.7	21.6	21.6		Filled-in	None	
13:15	POOH	3032.18	27	23.7	21.6	21.6			None	
13:30	Circulation	3032.18	27	23.7	21.3	21.6	-0.3		None	
13:45	Circulation	3032.18	26.9	23.7	21.2	21.6	-0.2	Consumed	None	

FIG. 11-6-1   A defective record table (A).

In Figure 11-6-2, there is no drilling fluid volume per tank, the drill string volume is discharged while tripping, and the drilling fluid is supplemented during normal drilling. There are the performance and flow rates of drilling fluid in the drilling tour report and drilling fluid tour report. These data do not have a direct relation with the variation of drilling fluid, and moreover, deleting these data is beneficial to reduce the operator's amount of work on duty. The observed result of sulfur hydrogen is generally in a normal state. It is not necessary to fill this blank if sulfur hydrogen is not monitored. Moreover, we can ignore this on the record table on duty.

Flow Monitoring And H₂S Observation Records

Well NO.   Supervisor On-duty:

min	Well Depth From	To	Drilling	POOH (String)	RIH (String)	Mud Properties Density	Viscosity	Flow Rate L/S	Total Surface Mud Volume m³	Change in Mud Volume m³	Well Condition	Concentration in Returned Mud (ppm)	Concentration in Tank (ppm)	Concentration in Pump House (ppm)	Observer
15:15	7158		Circulation			1.72	40	50	189	-0.3	Adding Water	0	0	0	
15:30	7158		Circulation			1.72	40	50	189	0	Normal	0	0	0	
15:45	7159		Circulation			1.32	34	50	189	0	Normal	0	0	0	
16:00	7158		Circulation			1.32	40	50	189	0	Normal	0	0	0	
16:15	7158		Circulation			1.32	40	50	189	0	Normal	0	0	0	
16:30	7158		Circulation			1.32	40	50	189	0	Normal	0	0	0	
16:45	3158		POOH	2					188.8	-0.2	Normal	0	0	0	
17:00	7158		POOH	2					188.6	-0.2	Normal	0	0	0	
17:15	3158		POOH	2					188.4	-0.2	Normal	0	0	0	
17:30	2158		POOH		2				188.6	+0.2	Normal	0	0	0	
17:45	7158		POOH		2				188.8	+0.2	Normal	0	0	0	
18:00	3158		POOH		2				189	+0.2	Normal	0	0	0	
18:15	3158		Circulation			1.32	42	50	189	0	Normal	0	0	0	

FIG. 11-6-2   A defective record table (B).

In Figure 11-6-3, there are no stand numbers of the trip, the drill string volume is discharged while tripping, and drilling fluid is supplemented during normal drilling; as a result, it is hard to analyze the reason for the drilling fluid's variation. The data of the liquid level is not intuitive and it is needed to be converted into volume, so this data should be deleted.

FLOW MONITORING RECORDS

Date			MM-DD-YY	Mud Density				Mud Viscosity						Supervisor On-duty	
Time h:min	Working Condition	Depth m	)#Tank Mud(Liquid Level cm	VOL.m³	)#Tank Mud(Liquid Level cm	VOL.m³	)#Tank Mud(Liquid Level cm	VOL.m³	)#Tank Mud(Liquid Level cm	VOL.m³	Accumulative VOL.	VOL. Increased or Decreased	Ditch Indication and analysis on increase or decrease	Person On-duty Signature
12:00														
12:15														
12:30														
12:45														

FIG. 11-6-3   A defective record table (C).

Remedy

The operator on duty should record the data of having a direct relation with the drilling fluid's variation on the same table. In this way, it is easy to obtain the variation of drilling fluid and easy to analyze the reason for the drilling fluid's variation. This kind of table should include the following content: date, recorder, cadre on duty, safety supervisor, time, operating mode, well depth, accumulative total of numbers of stands while tripping, the drill string volume discharged while tripping in theory, accumulative inject (return) volume, the drilling fluid's amount for each single tank, drilling fluid supplement, drilling fluid's variation, and the cause analysis (Fig. 11-6-4).

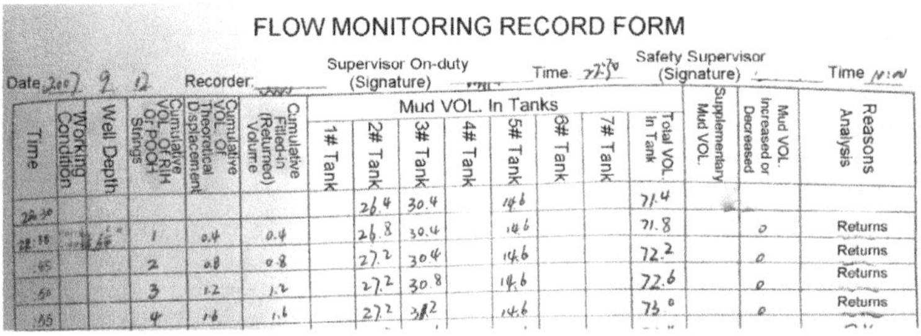

FLOW MONITORING RECORD FORM

Date 2007 9 12 Recorder: [illegible] Supervisor On-duty (Signature) [illegible] Time [illegible] Safety Supervisor (Signature) [illegible] Time [illegible]

Time	Working Condition	Well Depth	Cumulative VOL Of RIH Or POOH Strings	Cumulative VOL Of Theoretical Displacement	Cumulative Filled-in (Returned) Volume	Mud VOL. In Tanks 1# Tank	2# Tank	3# Tank	4# Tank	5# Tank	6# Tank	7# Tank	Total VOL. In Tank	Supplementary Mud VOL.	Mud VOL. Increased or Decreased	Reasons Analysis
[illegible]							26.4	30.4		14.6			71.4			
[illegible]	[illegible]	[illegible]	1	0.4	0.4		26.8	30.4		14.6			71.8		0	Returns
45			2	0.8	0.8		27.2	30.4		14.6			72.2		0	Returns
50			3	1.2	1.2		27.2	30.8		14.6			72.6		0	Returns
55			4	1.6	1.6		27.2	31.2		14.6			73.0		0	Returns

FIG. 11-6-4 Recommended record table on duty.

Hidden danger: The pressure gauge is installed horizontally

Hazard

If the pressure gauge is installed horizontally, the potential error of the pressure gauge's value is huge. This is not good for obtaining the exact pressure. It is easy to find a horizontally installed pressure gauge in a drilling site. The pressure gauge on the throttle and kill well manifolds and liquid-gas separator should not be installed horizontally (Figs. 11-6-5, 11-6-6).

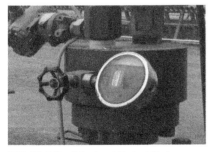

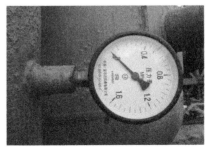

FIG. 11-6-5 A horizontally installed pressure gauge on the throttle manifold.

FIG. 11-6-6 A horizontally installed pressure gauge on the liquid-gas separator.

Remedy

The pressure gauge must be installed vertically instead of horizontally.

If the pressure gauge is installed horizontally, the pressure tap will be horizontal but there is a lack of axial type pressure gauge (T type; the connection caliber and dial plate compose a T shape) in the drilling crew. As a result, we can take the following two methods to avoid a horizontally installed pressure gauge: First, use a right angle bent sub between the horizontal pressure tap and radial type pressure gauge (I type; the connection caliber and dial plate compose an I shape; Fig. 11-6-7), just as Figures 11-6-8 and 11-6-9 show. The second method is to install an axial type pressure gauge on the pressure tap, as Figure 11-6-10 shows.

FIG. 11-6-7   A radial type (I type) pressure gauge.

FIG. 11-6-8   A vertically installed pressure gauge on the throttle manifold.

FIG. 11-6-9   A vertically installed pressure gauge on the liquid-gas separator.

FIG. 11-6-10   An axial type (T type) pressure gauge.

Chapter 12

Exercise: Identifying Hidden Dangers

This chapter can serve as a hazard identification exercise. Section 1 contains pictures in which there are hidden dangers; section 2 identifies the hidden dangers for each corresponding picture.

1. PICTURES OF HIDDEN DANGERS OR DEFECTS

Please identify the hidden dangers and defects in the following pictures. The answers are in section 2.

FIG. 12-1-1   FIG. 12-1-2

FIG. 12-1-3

FIG. 12-1-4

FIG. 12-1-5

FIG. 12-1-6

FIG. 12-1-7

FIG. 12-1-8

FIG. 12-1-9

FIG. 12-1-10

FIG. 12-1-11

FIG. 12-1-12

FIG. 12-1-13

FIG. 12-1-14

FIG. 12-1-15

FIG. 12-1-16

FIG. 12-1-17

FIG. 12-1-18

FIG. 12-1-19

FIG. 12-1-20

FIG. 12-1-21

FIG. 12-1-22

FIG. 12-1-23

FIG. 12-1-24

FIG. 12-1-25

FIG. 12-1-26

FIG. 12-1-27

FIG. 12-1-28

FIG. 12-1-29

FIG. 12-1-30

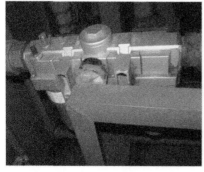

FIG. 12-1-31

FIG. 12-1-32

FIG. 12-1-33

FIG. 12-1-34

FIG. 12-1-35

FIG. 12-1-36

FIG. 12-1-37

FIG. 12-1-38

FIG. 12-1-39

FIG. 12-1-40

FIG. 12-1-41

FIG. 12-1-42

FIG. 12-1-43

FIG. 12-1-44

FIG. 12-1-45

FIG. 12-1-46

FIG. 12-1-47

FIG. 12-1-48

FIG. 12-1-49

FIG. 12-1-50

FIG. 12-1-51

FIG. 12-1-52

FIG. 12-1-53

FIG. 12-1-54

FIG. 12-1-55

FIG. 12-1-56

FIG. 12-1-57

FIG. 12-1-58

FIG. 12-1-59

FIG. 12-1-60

FIG. 12-1-61

FIG. 12-1-62

FIG. 12-1-63

FIG. 12-1-64

FIG. 12-1-65

FIG. 12-1-66

FIG. 12-1-67

FIG. 12-1-68

FIG. 12-1-69

FIG. 12-1-70

FIG. 12-1-71

FIG. 12-1-72

FIG. 12-1-73

FIG. 12-1-74

FIG. 12-1-75

FIG. 12-1-76

FIG. 12-1-77

FIG. 12-1-78

FIG. 12-1-79

FIG. 12-1-80

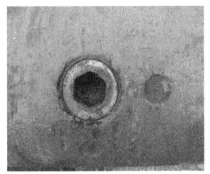

FIG. 12-1-81

FIG. 12-1-82

FIG. 12-1-83

FIG. 12-1-84

FIG. 12-1-85

FIG. 12-1-86

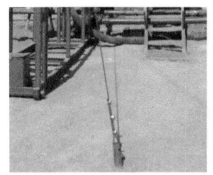

FIG. 12-1-87

FIG. 12-1-88

FIG. 12-1-89

FIG. 12-1-90

FIG. 12-1-91

FIG. 12-1-92

FIG. 12-1-93

FIG. 12-1-94

FIG. 12-1-95

FIG. 12-1-96

FIG. 12-1-97

FIG. 12-1-98

FIG. 12-1-99

FIG. 12-1-100

FIG. 12-1-101

FIG. 12-1-102

FIG. 12-1-103

FIG. 12-1-104

FIG. 12-1-105

FIG. 12-1-106

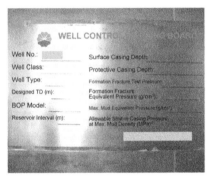

FIG. 12-1-107

FIG. 12-1-108

FIG. 12-1-109

FIG. 12-1-110

FIG. 12-1-111

CROSS REFERENCE OF RIH&POOH AND Change in Liquid Level of Metering Tank			
Drilling Strings		Drilling String Size m³	Change in Liquid Level of Metering Tank m³
127 mm	1 String	0.105	Half a small scale
	2 Strings	0.201	One small scale
	3 Strings	0.315	One and a half small scale
	4 Strings	0.42	Two small scales
177.8 mm	1 String	0.55	Two small scales
	2 Strings	1.1	One big scale

FIG. 12-1-112

When flow occurs while tripping

1. Give out signals and stop tripping.

2. Rush to connect the back pressure valve (or drop check valve in); Rush to run in the hole with drill pipes or make up Kelly (Drill rig with top drive: rush to connect the top drive head)

3. Open the choke valve properly.

4. While closing BOPs, annular BOP must be closed first, and then the pipe ram.

5. Close the choke valve to try to shut in the well. Rush to run in the hole with drill pipes or close the blind ram BOP.

6. Report to the drilling toolpusher and technicians promptly.

7. Record accurately standpipe pressure, casing pressure and mud volume change and get ready to shut in the well.

FIG. 12-1-113

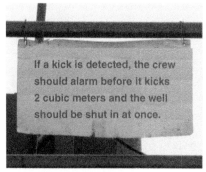

If a kick is detected, the crew should alarm before it kicks 2 cubic meters and the well should be shut in at once.

FIG. 12-1-114

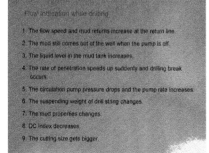

Flow indication while drilling

1. The flow speed and mud returns increase at the return line.

2. The mud still comes out of the well when the pump is off.

3. The liquid level in the mud tank increases.

4. The rate of penetration speeds up suddenly and drilling break occurs.

5. The circulation pump pressure drops and the pump rate increases.

6. The suspending weight of drill string changes.

7. The mud properties changes.

8. DC index decreases.

9. The cutting size gets bigger.

FIG. 12-1-115

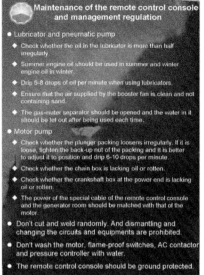

FIG. 12-1-116   FIG. 12-1-117

2.   ANSWERS

Fig. 12-1-1:

1. The BOP fixed guy lines are not arranged along the diagonal of the derrick substructure.
2. The mud blocking umbrella is poor, and cannot guarantee that the BOP and drilling units under the substructure are clean.
3. The gate valve wheel of the spool lacks operating platforms.

Fig. 12-1-2:

The guy line position is too low; guy lines are at the spool flange.

Fig. 12-1-3:

1. The mud blocking umbrella is too small; it cannot guarantee that the BOP and the area under the derrick substructure are clean.
2. The lower pipe-ram BOP manually operated rods are not securely connected; the manually operated lever with the BOP manual locking shaft is off.

Fig. 12-1-4:

1. The mud blocking umbrella declines, drilling fluid cannot be gathered, and the hygiene of the area under the derrick substructure cannot be guaranteed.
2. The cellar is a shortened operational platform.
3. The No. 3 gate valve hand wheel lacks an outer ring.

Fig. 12-1-5:

1. All manual operation rods are not connected outside the derrick substructure.
2. The chain fall block is used in the guy lines of the BOP.
3. The No. 4 gate valve is manual.

Fig. 12-1-6:

1. The BOP guy line diameter is less than 16 mm and be knotted, and the rope clamps used to fix guy line is less than 3.
2. Two lower pipe-ram BOP manual operation rods are not connected beside the derrick substructure.
3. The No. 4 gate valve is operated manually.

Fig. 12-1-7:

1. The manual operation rods are not connected outside the derrick substructure.
2. The manual operation rods are not connected with a cardan joint.

Fig. 12-1-8:

1. Valve 3 is a hydraulic valve.
2. Chain blocks are used in the BOP fixed guy lines.
3. The side outlet hole faces toward the drawworks (the hydraulic lines face the V-door).
4. There is poor sanitation under the derrick floor.

Fig. 12-1-9:

1. The operation platforms around the square well (cellar) are replaced with a guardrail.
2. The angle of the manual operation rod on the right upper ram BOP is greater than 30°.

Fig. 12-1-10:

1. The BOP fixed guy lines are fixed on annular BOP bolts.
2. A fixed guy line is knotted.

Fig. 12-1-11:

1. Guy lines are fixed on the BOP side door bolts.
2. The BOP guy lines are not fixed along the diagonals of the derrick floor.

Fig. 12-1-12:

One side of the propping bar is fixed to BOP flange bolts.

Fig. 12-1-13:

1. The manually operated rods of the upper ram BOP are not connected outside the derrick substructure.

2. It is not easy to operate because the hand wheel of the upper ram BOP is influenced by the rail fence.
3. There are no handrails on the inclined ladder on the upper operation platform, and it is wrong to use stick stairs instead of pedal stairs.

Fig. 12-1-14:

The manual operation rod of the upper ram BOP is connected outside the derrick substructure.

Fig. 12-1-15:

1. There is no operation platform on the cellar.
2. The manual operation rod is not connected outside the derrick substructure.
3. There is no outer ring of the manual operation rod hand wheel.

Fig. 12-1-16:

1. The square well/cellar operating platform is not fully paved, and there are holes.
2. The manual operation rod of the lower ram BOP is not connected outside the derrick substructure.
3. The BOP fixed guy line only uses one rope clamp.
4. The BOP is painted in blue.

Fig. 12-1-17:

1. The manual lever operation platform has no guardrail; there are no handrails on the inclined ladder.
2. The shut-in sign tag/band does not face the operator.

Fig. 12-1-18:

The flange bolt standard is not unified.

Fig. 12-1-19:

The hydraulic lines and the inlet of the annular BOP are not connected with a bent sub.

Fig. 12-1-20:

The hydraulic valve is connected to the inside of the manual valve.

Fig. 12-1-21:

The hydraulic lines and the inlet of the ram BOP are not connected with a bent sub.

Fig. 12-1-22:

The blind-ram BOP uses the position restricted device.

Fig. 12-1-23:

The handle of the blind-ram BOP reversing valve is bounded with wire to restrict its position.

Fig. 12-1-24:

The reversing valve of the blind-ram BOP is in a restricted position with a pin.

Fig. 12-1-25:

The damper oil of the anti-seismic pressure gauge leaks.

Fig. 12-1-26:

Pneumatic pumps are not connected into the gas pipeline.

Fig. 12-1-27:

The visual oil cap of the pneumatic pump is damaged, and the wire plugs drip oil.

Fig. 12-1-28:

1. The hydraulic lines' self-sealing quick coupler leaks oil.
2. The pipe joints do not have anti-pollution measures.

Fig. 12-1-29:

The annular blowout preventer pressure is below 10.5 MPa.

Fig. 12-1-30:

The remote console hydraulic oil spills, and there is no way to staunch the flow.

Fig. 12-1-31:

1. The pneumatic pump barometer dial is covered with paint, and does not recognize the numbers.
2. The barometer is obstructed by a bracket, making it difficult to read the data and replace the barometer.

Fig. 12-1-32:

1. Flange surfaces are not parallel.
2. The flange nuts are not tightened.
3. The back-up torque of bolts is not the same; one side of the nut has surplus threads, and another side reaches too long.

Fig. 12-1-33:

The gate valve screw stem bends.

Fig. 12-1-34:

The BOP hydraulic lines block the valve hand wheel rotation.

Fig. 12-1-35:

1. The gate valve hand wheel is not standard and rotates inconveniently.
2. The cellar does not have a platform.
3. The spool bottom flange bolts are not up to standard; the bolts are too long.

Fig. 12-1-36:

The hydraulic valve's lines hamper hand wheel rotation.

Fig. 12-1-37:

Valve 3 is a hydraulic gate valve.

Fig. 12-1-38:

The length of flare line is more than 7 m. It is not fixed.

Fig. 12-1-39:

The flare line is not fixed at the corner.

Fig. 12-1-40:

1. The wrong color tag for the valve is hung; green means pass, and red means closed.
2. The relief pipe and liquid feed pipe elbows of the liquid-gas separator are not fixed.
3. The liquid-gas separator liquid feed line and relief line colors are yellow or green.
4. The killing well sign lacks the maximum allowable shut-in casing pressure under different drilling fluid densities.
5. There is no low surveying range pressure gauge.

Fig. 12-1-41:

1. A nonstandard choke manifold is used.
2. There is no buffer tube; there are no J5, J7, J8, and J9 gate valves.

Fig. 12-1-42:

The nut has surplus threads.

Fig. 12-1-43:

The pressure gauge installation is tilted (over 30°).

Fig. 12-1-44:

There is no tag on the pressure gauge of the chock manifold.

Fig. 12-1-45:

There is no tag on the pressure gauge of the kill manifold.

Fig. 12-1-46:

1. The outlet of the drilling fluid recovery pipeline is fixed at the catwalk.
2. The steel tubes are welded directly to make the bent sub of the drilling fluid recovery line.

Fig. 12-1-47:

The hand wheel of the manual choke valve is loosening, causing failure.

Fig. 12-1-48:

The drilling fluid recovery line uses a steel tube welding bent sub.

Fig. 12-1-49:

The bent sub is field welded with a steel tube as a recovery pipeline; the bent sub is in the air.

Fig. 12-1-50:

The recovery pipe is welded in the field and dislocated.

Fig. 12-1-51:

1. The recovery pipe is not fixed at the corner.
2. A curved steel pipe was used instead of a cast steel bent sub.

Fig. 12-1-52:

1. The recovery line uses a noncast steel bent sub.
2. The bent sub outlet faces reinforcing steel, resulting in poor drainage.

Fig. 12-1-53:

The fixed gland of the recovery hose of the drilling fluid at the upper tank is not fixed tightly.

Fig. 12-1-54:

The drilling fluid recovery hose line is too long, and the middle is not fixed.

Fig. 12-1-55:

The outlet of the recovery pipe is welded directly to the tank edge and is not solid.

Fig. 12-1-56:

The recovery line is too long (more than 10 m), and it is not fixed.

Fig. 12-1-57:

1. The recovery steel pipe outlet uses a nonsteel bent sub.
2. Wooden plugs are used under the fixed gland/plate.
3. The outlet fixed-point is welded directly to the tank edge and is not solid.

Fig. 12-1-58:

1. The corner of the relief pipe is not fixed.
2. The relief pipe is painted yellow.

Fig. 12-1-59:

1. The recovery line is used with a union.
2. The corner is not fixed.

Fig. 12-1-60:

The blowout pipe interface is located under the pipe bridge; it is not easy to check and maintain.

Fig. 12-1-61:

The blowout pipeline length is too short, less than 20 m.

Fig. 12-1-62:

The corner is not fixed.

Fig. 12-1-63:

The blowout pipeline pier base is near the edge of the waste pit; the foundation is not solid.

Fig. 12-1-64:

There is no blowout pipeline bridge.

Fig. 12-1-65:

1. The blowout pipe uses a union (by Wang) connection.
2. The blowout pipeline uses on-site welding.

Fig. 12-1-66:

1. The length of the blowout pipeline is more than 15 m and not fixed.
2. The flare line is through the waste pit.

Fig. 12-1-67:

1. A fixed base pier side door is leaking.
2. The fixed base pier doesn't have an upper cover.
   The casing is used as a blowout pipe.

Fig. 12-1-68:

The fixed base pier side door is not closed tightly, which will cause filler to leak (the vibration is more severe) and reduce the weight of the base pier.

Fig. 12-1-69:

There are surplus threads at the relief pipe bent sub thread connection.

Fig. 12-1-70:

1. The relief pipeline bridge is not protecting the pipeline.
2. The color of the blowout is yellow.

Fig. 12-1-71:

The relief pipe is solidified into the anchor by cement, so that the replacement or displacement is difficult.

Fig. 12-1-72:

The relief pipeline bent sub is solidified into the anchor by cement, so that the replacement or displacement is difficult.

Fig. 12-1-73:

1. The relief pipeline outlet is fixed by a single base pier.
2. The fixed-point is too far away from the outlet, about 4 m.
3. The drill pipe threads outward.

Fig. 12-1-74:

The relief pipe faces hydraulic lines.

Fig. 12-1-75:

There are flammable crops near the outlet of the relief pipeline, nonblocking firewall, or the combustion chamber at the outlet.

Fig. 12-1-76:

1. The outlet of the relief pipe is located at the entrance door of the well site.
2. The emergency assembly point is near the relief pipeline outlet.
3. There is no relief pipeline bridge.

Fig. 12-1-77:

1. The weight of the base pier is less than 500 kg.
2. The pipeline fixed gland is not fixed.
3. The inside threads outward.

Fig. 12-1-78:

1. The length of the relief pipeline is too short, about 30 m.
2. The distance between the outlet of the relief pipeline and logging room is less than 8 m.

Fig. 12-1-79:

1. A fixed base pier can be closer to the bent sub.
2. The drilling fluid recovery hose is too long and not fixed.

Fig. 12-1-80:

The relief pipe bolts are welded to the shale shaker base; this approach is not firmly fixed.

Fig. 12-1-81:

The cock is in a semi-open state.

Fig. 12-1-82:

When tripping drill pipe, the BOP single is put on the ramp and stretched out of the derrick floor; there is the risk of injury to employees.

Fig. 12-1-83:

The cock switch label is ambiguous.

Fig. 12-1-84:

All the switching instructions are off.

Fig. 12-1-85:

The cock handle water course is blocked.

Fig. 12-1-86:

The drilling check valve is seriously corroded.

Fig. 12-1-87:

The guy lines lack steamboat ratchets.

Fig. 12-1-88:

The liquid feed pipe of the liquid-gas separator uses an asbestos sealing flange.

Fig. 12-1-89:

The liquid-gas separator gas relief pipe's diameter is decreased.

Fig. 12-1-90:

1. The liquid feed hose of the liquid-gas separator is not fixed.
2. The hose has not used the safety chain.

Fig. 12-1-91:

1. The liquid feed hose of the liquid-gas separator is connected by a union.
2. The liquid-gas separator is painted blue.

Fig. 12-1-92:

1. The inlet of the liquid-gas separator discharge line is lower than the outlet.
2. The safety valve pressure relief port is connected with a slender bent discharge pipe.
3. The liquid-gas separator is painted blue.

Fig. 12-1-93:

1. The liquid-gas separator pressure gauge is connected to the buffer tube of the liquid feed pipe.
2. The gas relief pipe's diameter is decreased.
3. The liquid-gas separator is painted blue.

Fig. 12-1-94:

The outlet of the liquid-gas separator discharge line is lower than the inlet.

Fig. 12-1-95:

The pressure relief valve discharge pipe of the liquid-gas separator is connected with the gas relief pipe port; release is not possible.

Fig. 12-1-96:

The pressure gauge has a 90° tilted installation; here you should not use an I-type pressure gauge.

Fig. 12-1-97:

The pressure gauge dial faces the ground; here you should not use an I-type pressure gauge.

Fig. 12-1-98:

The U-clamp fixed guy line is stuck in the main line.

Fig. 12-1-99:

The outlet of the liquid-gas separator discharge line is lower than the inlet position.

Fig. 12-1-100:

The outlet of the liquid-gas separator discharge line is as high as the height of the inlet.

Fig. 12-1-101:

1. The liquid-gas separator's inlet/outlet pipe is welded.
2. The liquid feed pipe is connected by a union.
3. The liquid feed pipe of the liquid-gas separator is connected from the ninth gate valve.

Fig. 12-1-102:

The liquid-gas separator rope clamp interval is about 50 cm, more than 20 cm.

Fig. 12-1-103:

The pressure relief port of the liquid-gas separator safety valve is connected with a curved slender discharge tube.

Fig. 12-1-104:

1. The liquid-gas separator guy lines are fixed in the middle, and they are not firmly fixed.
2. The liquid-gas separator is painted blue.

Fig. 12-1-105:

The degasser of the gas relief line is bent, and the passageway is blocked.

Fig. 12-1-106:

This does not indicate the current drilling fluid density corresponding to the maximum allowable shut-in casing pressure value.

Fig. 12-1-107:

This does not indicate the current drilling fluid density corresponding to the maximum allowable shut-in casing pressure.

Fig. 12-1-108:

The fire extinguisher is placed outdoors.

Fig. 12-1-109:

The second and third diesel engines don't have a cooling device and spark-free device.

Fig. 12-1-110:

The pole lamp of the well site area is not explosion-proof.

Fig. 12-1-111:

The explosion-proof electrical control box on the drilling fluid tank lacks fastening bolts, which is an explosion-proof performance failure.

Fig. 12-1-112:

1. Changes in the amount of drilling fluid volume are accurate to three decimal places, are not absolutely necessary, and create an unnecessary workload.
2. The volume changes in one small scale multiple 2 is not equal to the volume changes of the two small scales.

Fig. 12-1-113:

1. There are shut-in program errors.
2. The second procedure: After rushing to connect the back-pressure valve during tripping drill pipe, rush to run in the hole with a drill pipe or make up a Kelly without closing the ram BOP.
3. The fifth procedure: After shut-in, still open the ram BOP to run in the hole with drill pipe and close the blind-ram BOP with drill tools in the hole.

Fig. 12-1-114:

A kick is detected and the well is not shut in at once and the kick develops into 2 cubic meters.

Fig. 12-1-115:

The overflow may not happen even under situations such as:
- drilling fluid tank liquid level rise
- rate of penetration speeds up suddenly
- emptying
- circulating pump pressure drops
- drill string free hanging weight changes
- changes in drilling fluid performance
- Dc index decreases and debris size becomes larger

In other words, these situations can also happen even if no overflow occurs.

Fig. 12-1-116:

The unit of scale ruler mark is surplus and interferes with the on-duty operator's line of sight.

Fig. 12-1-117:

There are many tags on the equipment reading inspecting equipment "periodically", which does not define the definite time period and lacks of operability.

Index

Note: Page numbers followed by '*f*' indicate figures.

Printed and bound by CPI Group (UK) Ltd, Croydon, CR0 4YY

03/10/2024

01040416-0003